The
Office Safety
Handbook

HEALTH & SAFETY IN PRACTICE

The Office Safety Handbook

Roger Saunders

PITMAN PUBLISHING

PITMAN PUBLISHING
128 Long Acre, London WC2E 9AN

A Division of Pearson Professional Limited

© Roger Saunders 1995

First published in Great Britain, 1995

British Library Cataloguing in Publication Data
A CIP catalogue record for this book can be obtained from the British Library.

ISBN 0 273 61246 8

All rights reserved; no part of this publication may be reproduced, stored in a retrieval system, or transmitted in any form or by any means, electronic, mechanical, photocopying, recording, or otherwise without either the prior written permission of the Publishers or a licence permitting restricted copying in the United Kingdom issued by the Copyright Licensing Agency Ltd, 90 Tottenham Court Road, London W1P 9HE. This book may not be lent, resold, hired out or otherwise disposed of by way of trade in any form of binding or cover other than that in which it is published, without the prior consent of the Publishers.

10 9 8 7 6 5 4 3 2

Typeset by Northern Phototypesetting Co Ltd., Bolton
Printed and bound in Great Britain by
Bell & Bain Ltd., Glasgow

The Publishers' policy is to use paper manufactured from sustainable forests

Contents

Preface vii

Acknowledgements ix

❶ Your role as office manager 1
Introduction   1
Implications of HSWA   2
Information analysis   5
Remedial actions   5
Safety programmes   6
Action checklist   21

❷ Legal requirements 23
Introduction   23
Historical background   24
Employer liability   27
Criminal liability   31
Health and Safety at Work Act 1974   40
Enforcement   43
Personal liability   46
Acts of Parliament   51
Regulations   56
The European Union   61
Summary   61
Action checklist   63

❸ How to implement your safety policy 65
Safety policy statement   65
Safety representatives and committees   67
Accidents and dangerous occurrences   70
Liability insurance   73
Women workers   74
Industrial tribunals   74
Action checklist   78

4 Accident records 81
Information gathering 81
Data sources 88
Primary and secondary information 89
Accident data collection 92
Employee co-operation 99
Action checklist 105

5 What to look for in the office 107
Fire 107
First-aid 112
Heating 114
Lighting 115
Ventilation 116
Noise and vibration 118
Radiation – VDUs 119
Protective clothing and equipment 121
Manual handling 123
COSHH 125
Welfare 128
Alcohol and drugs 129
Stress management 129
Cleaning 131
Hygiene 132
Action checklist 132

6 Finance 135
Introduction 135
Accident costing 136
Budgetary control 140
Action checklist 143

7 Electricity and machinery 145
Electrical equipment 145
Maintenance procedures 145
Action checklist 148

8 Auditing safety 151
What is safety auditing? 151
Safety policies 155
Auditing safety procedures 155
Auditing safety practices 156
Auditing safety programmes 156
Safety audit management 158
Action checklist 164

Useful organisations 166
Further reading 169
Index 171

Preface

For the purposes of writing this book an office is taken to mean any room or premises where five or more persons work primarily on administrative matters. However, this does not mean that the book will not be of interest to those people who work in offices with less than five employees. On the contrary, this book is a source of information for all section heads and office managers whether they operate on their own or as part of a team.

The office is usually regarded as the administrative core of an organisation. In some companies managers have often asked the question: who works for whom? Does management support the administration or *vice versa*? Most important matters are usually dealt with ultimately by the *office*. In larger organisations, the office is usually responsible for discipline, policy, procedures, practice, pay and so on whereas in smaller companies the office may be all that exists and everything takes place within its walls. Health and safety applies to us all whether we are the management or individuals working alone.

Effective office safety and health policies and practice emanate from three fundamental factors:

- the building itself
- the people working in the building
- the equipment, furnishings and fittings within it.

For example, on one recent visit to a well-known local newspaper, it was found that three journalists (one fairly well-known for writing on health and safety matters) occupied an office or newsroom which contained:

- three computers
- one laser printer
- one kettle
- one bar heater.

From another electrical point:

- a fax machine
- one photocopying machine
- one coffee percolator
- one typewriter.

All this was supplied from two extension boards plugged into each socket. Casual observation showed the socket to be of the old 5 amp round pin type.

The building was a Victorian conversion, but the electric supply could be seen to be rather old and had obviously not been updated when the house was converted for office use. At the same time, electrical cabling ran all over the place. This office was not unique. I have seen many dangerous offices such as this, from solicitors' offices to supermarket offices. My personal experience has found these to be particularly evident in cities where old houses have been converted into office accommodation.

Although the building described above was old, those responsible for the original office selection should have considered the three fundamental factors mentioned above. The adequacy of the electrical system is an obvious one to consider. A section is included in this book to cover the basic things to look for when considering the purchase of rental of office accommodation.

Although the *office* is not described as a place for vigorous active pursuits it has nevertheless proven to be an extremely dangerous place to work if simple rules are not adhered to. Examination of office accident records and personnel files from several companies over the past five years show the office to be as dangerous as the shop floor or construction site. Fatalities occur regularly as do serious injuries. In these specific cases, they occurred from falls from ladders whilst *filing* or having a filing cabinet or book shelf topple on top of the person (and in two cases persons). Minor injuries abound from cut fingers, bruising, sprains, burns and believe it or not, stapled fingers! Further health problems were found from sickness records. Headaches caused by poor ventilation (using propane type heaters and poorly ventilated gas water heaters) and stomach problems caused by poor washing facilities where staff were found using the only wash-hand basin (which was broken) for cleaning cups and plates and hands after going to the toilet.

Absences through injury and sickness were known to the companies mentioned above, but it was only when those costs involved were spelt out to the managers concerned did they decide that health and safety, rather than originally being seen as an expense, has in fact now shown to be a clear saving.

This book has been primarily written to provide office managers and their staff with some simple, inexpensive information which should assist them to develop and implement measures to reduce and/or prevent accidents from happening, thus reducing company costs.

Dr R A Saunders
1995

Roger Saunders has over 20 years experience as a safety practitioner and specialises in safety management. He has published widely and has undertaken several projects for the World Health Organisation, United Nations, World Bank and Foreign Governments. He also has wide project experience in UK industries, commerce and local government.

Previously he was Director of Research at the Centre for Safety Studies and until 1992, was Director of Studies for the MSc in Safety Management Studies at Bournemouth University.

Acknowledgements

In order to write a book of this nature it is necessary to have the assistance of others if the information contained in the book is to be technically sound and in this regard I am indebted to the library staffs at RoSPA and the British Safety Council for their kind help which they afforded to me throughout this project. Some of the information was required whilst I was working overseas but this was not a deterrent as far as the staff were concerned. I was tracked down wherever I was. In addition, I would like to thank the staff of Dorsch Consult's Safety Management Group and in particular Dipl. Ing. H Pihan for providing some of the practical illustrations and checklists which have been included in the main text.

In addition, I would like to thank those many companies which have provided information concerning good office safety principles, practices and management and given permission for it to be cited in the text.

Finally, I must thank my friends and loved ones for their moral support and kindness throughout the long hours spent in researching appropriate material and in typing the text and to my dear and young Patrice for allowing me the time to do the work necessary to meet publishing deadlines.

'Your major areas of responsibility are the implementation and monitoring of company health and safety policy'

'Following the introduction of HSWA 1974, organisations have had a statutory obligation to ensure the safety and well-being of their employees'

Your role as office manager

INTRODUCTION

All office managers or supervisors charged with the responsibility of carrying out an organisation's health and safety policy must do it efficiently and effectively. To do this, it is important you are aware of your duties, responsibilities and functions. To assist in this it is important to remember that there are eight basic office safety activities to consider. By being responsible for health and safety policy, procedure and practice in your particular workplace you should consider the following key task areas:

- you will need to *plan* policies, procedures and programmes designed to reduce accidents or to prevent them from happening. Accident prevention also means the inclusion of all health matters
- you will need to regularly *monitor* all health and safety policies, procedures, practices and programmes to ensure satisfaction
- you will be required to provide on the job *instruction* and *training* in the following:
  - safe systems of working
  - workplace rules and regulations
  - employee responsibilities
  - indoctrination and education
- you will need to act as an accident *information gatherer*
- you will be required to *provide* and *issue* appropriate protective clothing and or equipment in accordance with the law and appropriate regulations
- you will also be required to *maintain* a safe environment in which you expect people to work. If you are in any doubt whatsoever about your duties and responsibilities you should know when and how to seek help
- you should get to *know* and *understand* your staff. Do not expect an employee to do a job they know nothing about and you should be seen to lead by example

- you have a duty to *ensure that health and safety matters are given a high priority*. It is important to secure recognition of the importance of health and safety at senior level and sufficient resources should be allocated to the tasks.

These activities are summarised in Figure 1.1 opposite.

Your major areas of responsibility are the implementation and monitoring of company health and safety policy. You will usually have to manage other staff and be faced with the planning, organisation, motivation and control of them. Following the introduction of the Health and Safety at Work Act (HSWA) in 1974, organisations have had a statutory obligation to ensure the safety and well-being of their employees. This is a general piece of legislation covering broad safety principles whilst other forms of legislation such as the Factories Act 1961 or the Fire Precautions Act 1971 are more specific (see Chapter 2). Some organisations react more enthusiastically than others in implementing such legislation. In this chapter, therefore, it is necessary to consider the implications of the HSWA which placed a broad statutory responsibility on employers to ensure the health, safety and welfare of their employees at work and discuss the daily operational management plan required to deal with this responsibility.

IMPLICATIONS OF HSWA

Information gathering

You will need to gather accident information so that your effective health and safety decisions can be made, and even with scarce resources accident reduction and/or prevention policies can be applied effectively. There are a number of sources of information which are useful when planning, implementing, monitoring and evaluating health and safety policies, programmes and procedures. Good sources of information regarding what is happening in the workplace can be gleaned from:

- internal sources such as:
  - monitoring and evaluation (safety audits)
  - accident and dangerous occurrence investigation
  - personnel records
- external sources such as:
  - professional journals
  - HSE circulars
  - newspapers.

Many companies monitor or evaluate their health and safety success (or failures) via a number of commercially available *safety auditing* systems whilst others opt for in-house methods. These systems are very useful for obtaining information to provide a safety rating or score. Very few firms base their strategies upon accident investigation and analysis techniques, and are therefore unable to show how effective they have been at reducing accidents

YOUR ROLE AS OFFICE MANAGER | 3

Task	Accident Investigator
Specific Activity	To carry out investigations into all accidents and dangerous occurrences in order to establish contributory factors.

Task	Advocate
Specific Activity	Establish health and safety as a priority within the organisation and secure its recognition at board level. Secure sufficient resources.

Task	Auditor
Specific Activity	Carry out regular examinations of current health and safety policy, procedures, practice and programmes to ensure satisfaction.

Task	Leader
Specific Activity	Know and understand the workforce and lead by example. Motivate workers and develop schemes and plans to change attitude and behaviour.

Task	Planner
Specific Activity	Plan, implement, monitor and evaluate remedial measures designed to reduce or prevent accidents from happening.

Task	Provider
Specific Activity	Issue protective clothing and/or equipment. A knowledge of the legal requirements is necessary. Seek expert help where necessary.

Task	Trainer
Specific Activity	Provide on–the–job training, safe systems of working, indoctrination, workplace rules and regulations, and employee responsibilities.

Figure 1.1 General tasks for the office manager

or dangerous occurrences from year to year. Because of this, some organisations experience difficulties in convincing senior management, such as the company accountant, to invest properly in health and safety. It is important, therefore, to have a good accident investigation and analysis programme such as that discussed in Chapter 4. This will greatly assist with the provision of appropriate justifications which are required.

For the planning of remedial action, you will need to have access to various other types of information such as:

- the number of injury accidents which have occurred
- the number of dangerous occurrences which have taken place
- a detailed breakdown of each incident in sufficient detail that will provide for the identification of cause and/or contributory factors
- a sufficiently large enough data base to allow for the identification of trends (time will provide this)
- supplementary information concerning accidents and dangerous occurrences from similar industries and environments
- relevant national statistics where they are available.

Although these points are dealt with in greater detail in Chapter 4, it is important to realise that this type of information is a vital part of the decision making process and provide for efficiency and effectiveness in practice. Generally, you would need to collect detailed information concerning all dangerous occurrences in their respective areas of responsibility. This must include details of injury accidents.

Previous studies have shown the poor quality of primary accident data collection procedures and that maximum use of the information collected is not made. Despite this, many of those charged with the responsibility for health and safety matters, continue to ignore or realise the value of a good reliable primary data source. A strategic objective here, therefore, must include data collection and to ensure that these data are used for analysis purposes. This must form part of the decision making process.

To do this within practical economic constraints, you should establish a system for collecting details of accidents and dangerous occurrences as outlined in Chapter 4. It is known that many organisations spend considerable sums of money in order to meet a potentially dangerous situation. If an objective is to collect all accident information, then steps should be taken to collect this data in a sufficiently detailed form so that potential problem areas can be identified. This will enable priorities to be set and ensure that resources are used effectively. At the same time, supplementary data should be systematically obtained in detailed structured interviews with casualties so that lessons can be learnt. Wherever possible, these interviews should be in total confidence and without fear of any consequent disciplinary action.

Other forms of information should be considered. It has become established practice for the office manager to be primarily involved in only certain areas of health and safety responsibility such as sending staff for agreed safety training whilst someone else is supposed to concentrate upon the other issues. Such separation should be avoided wherever possible. To enable organisational missions to be met at all, and for efficiency and effectiveness to be improved, both systems should work closely together as each of them plan and implement strategies using the same data base for decision making. Both are equally dependent upon the quality of these data and there should be a shared and co-ordinated approach to remedying any problems of unreliability. It is from this premise that other forms of relevant information can be systematically gathered for programme planning and evaluation purposes. It is not intended to delve too deeply into what additional information is necessary for a proactive office management style in this chapter. Information gathering, therefore, is of great importance and should be given high priority within the objective setting sequence.

INFORMATION ANALYSIS

The majority of office managers who are responsible for health and safety do not have direct access to any accident data base for decision making purposes. In any case, few have received training in its use. An unacceptably high number are unaware of its value and most have not considered whether their accident trends were a problem or not. Some office managers and/or supervisors collect accident data for establishing trends but tend to analyze individual cases rather than establish causes. Furthermore, office managers do not regularly monitor dangerous occurrences nor are they familiar with statistical packages or research methodology.

This is despite government departments, the HSE, RoSPA and BSC, regularly encouraging managers to evaluate their work. Few use basic research tools such as computers. The problem can be compounded because some organisations pursue staff recruitment policies that are not designed to address this problem. In too many cases, they fail to recruit numerate graduates for key management roles or to provide for adequate staff development in the areas of information technology and statistics.

REMEDIAL ACTIONS

Planning remedial actions can only effectively take place after all relevant information has been thoroughly investigated and analyzed. Central to the discharge of your duty is the planning of appropriate remedial action. Education, training and publicity should be the main issues within the office. This does not imply that the mere writing of an office memo on a health and safety subject is an effective use of time and resources.

It is now accepted that for safety education to be effective it must be continuous and should, therefore, be part of the daily management plan. Clearly the delegation of this responsibility to all staff is necessary. In some instances, qualified office managers are not appointed in large enough numbers solely to discharge their duties and responsibilities. In very large offices, the role of the office safety representative is usually seen as that of adviser, supervising the implementation of company policy, arranging for appropriate safety training and trying to ensure that safety is given an appropriate high priority on the office agenda.

SAFETY PROGRAMMES

Implementation

The implementation of operational objectives needs careful thought in two regards; first, the staff resource, and second, the financial resource. The accident data base discussed earlier has to be the prime means of determining priorities. One cannot expect a company to throw blindly scarce resources at everything and anything in an effort to be seen to be doing something when ignoring known accident high risk locations and/or activities.

At the end of each year or time period when data is available together with that for the previous two years, or period data, trends can be established and plans of action can be determined for the forthcoming year. The start of the new year or period is an ideal time to be looking at objective setting in a company environment adopting a fiscal year or quarterly period. You need to be aware of your organisation's budget-making policies. A large number of office safety managers/supervisors are not setting objectives or planning from any data base, let alone considering operational objective setting as a process. Programmes based upon historical activities and precedent seem to be the norm rather than planning in the light of changing operational requirements. The structured method proposed leads to a series of specific actions and proposals that you should seek to confer with colleagues in other offices within the organisation.

Monitoring

The monitoring phase is very important and there are two specific areas involved:

- systems management
- operations management.

Systems management
There must be an orderly and regular review of objectives and their implementation. Monitoring specific objectives at critical periods is also important. Fluctuating conditions may influence schemes, such as moratoriums on bud-

gets. If these issues are allowed to go unnoticed or unchecked then problems inevitably arise. Problems such as printing deadlines can mean a lost opportunity for example, whilst insufficient staff can affect other issues. Resource systems, therefore, need systematic scrutiny if programmes are to succeed. Shortfalls in the resource base should be identified at an early stage in the decision-making process and their implications should be clearly understood. All too often office managers can identify what they can do, but are unable to describe what they cannot do or list accurately why they cannot do it. It is assumed that safety education and training should be made available to all employees without favour or bias. Office management should know whether this is happening and should develop a system whereby shortfalls in organisational expectations can be clearly and easily identified and other priorities set. To quote numbers without qualification is not good management practice.

Operational systems

These are important if projects are to meet their aims and objectives. Health and safety training schemes need to be regularly monitored if safety standards are to be maintained and the operational divisions of the organisations need to be contacted regularly if they are to carry out their programmes consistently and effectively. For example, if an objective is to train each and every member of your office staff in a particular area once a year, then this can easily be monitored. If this objective has not been reached then evaluation is needed to show the reasons for this failure. Likewise, if a scheme has been implemented then each stage needs discussion and regular meetings need to take into account any tactical consideration. The majority of office managers do not keep accurate records of their activities. From an operational systems standpoint, information detailing each staff meeting with dates, times, personnel present, etc., has immense operational value. In a similar manner, details of other schemes and plans must be recorded. Many professional safety practitioners now readily accept that they do not know the full extent of dangerous activities being carried out in their organisations and are equally unaware how many dangerous incidents have occurred. If the management system does not seek to quantify this information at the outset then retrospectively-based action plans need to be considered. If a stated objective of the office safety programme is to acquire this information then it has to be made possible to obtain it.

Evaluation

In terms of evaluation, there are many commercially available safety audit systems which allow office managers to pursue safety policies which are capable of being evaluated. Simple safety programme evaluation is a subject which is dealt with later in the final chapter. However, in some organisations the responsibility for these, particularly the more comprehensive type, will fall directly with the company safety manager who is qualified to implement these.

Although the RoSPA, HSE and BSC actively encourage use of all such programmes, it is wrong that 'evaluation' does not feature on the majority of the current training courses designed for those managers responsible for health and safety. Moreover, research methodology as an operational tool is similarly lacking from these health and safety courses. The use of statistics does feature in a handful of courses but these are usually academic in nature. In the accident investigation and analysis field there are even fewer opportunities to receive training. At the more basic training level, statistical methods are not covered in sufficient depth, neither are relevant examples used to illustrate their use. The use of information technology and other aids to successful evaluation are sadly lacking in practice in most organisations.

Evaluation can only occur within a framework of clearly defined aims and objectives and if undertaken by properly trained staff. Many safety professionals argue that safety activities cannot be evaluated, for example because you cannot quantify how many deaths or injuries have been avoided. This is symptomatic of this fundamental issue. Surely, if a particular plan were introduced to encourage the use of protective clothing in a specific location, then its results could easily be measured. If, after the campaign, more people are found to be wearing the clothing than before, then some degree of success may have been achieved particularly if no such change occurred in the control group. Other areas of behaviour and attitude can be similarly evaluated. Likewise, management strategies can also be evaluated for efficiency and effectiveness. It is recognised that specific reductions in accidents directly attributable to a safety education, training and/or publicity programme are very difficult to establish in terms of a simple causal relationship. Nevertheless, the safety manager's contribution to the overall corporate approach in accident reduction is now acknowledged and the long-term value is becoming accepted. For example, a comparison of the number of accidents in relation to the vehicle population in this country would confirm this, particularly when compared with our EU partners. The office manager, therefore, can be an important and vital part of this corporate plan.

Managing safety

The role of the office manager/supervisor in relation to responsibilities has so far been reviewed within the context of the HSWA. Some organisations have reacted more enthusiastically to this legislation than others. As a result, there is no standardised approach to accident reduction in this country. There also seems to be no standard recruiting policy or training provision. It seems that safety is regarded by some as a backwater in the organisation and many firms have been found to put some 'problem' staff into safety roles; and then the age of many safety practitioners is too old to allow for substantial retraining.

Organisations must, therefore, consider a management style that is not just public relations or law enforcement orientated but actually geared to accident prevention and reduction. This demands a different portfolio of skills.

YOUR ROLE AS OFFICE MANAGER | 9

If those responsible for health and safety are to be appointed and trained for the next century, then is it appropriate to appoint people who are too old to undergo sustained professional retraining when compared to other younger professionals, and further, is it wise to appoint people whose personal abilities, experience and attributes are not necessarily best suited to this matter?

GATHER INFORMATION

- Collect:
  - accident data
  - reports and journals
  - newspaper articles
- Analyse accident information
- Attend seminars/conferences
- Gather experiences from colleagues
- Obtain research reports

Keep up to date

ANALYSIS OF DATA

- Carry out statistical analyses of accident data
- Produce abstracts of all other relevant data
- Place accident data into context with other information
- Rank accident data for priority action

Prepare information in simple, understandable form

ACTION PLANS

- Prepare the safety mix
- Consider the resource requirements
  - implementation and time scales
  - monitoring procedures
  - evaluation phases

Consider the effectiveness of your plans

Figure 1.2   The role of the office manager

Could this be the reason why those courses currently available are not well attended or is this because office managers responsible for training do not really know or understand what the role of the safety professional should be?

Several safety practitioners, particularly in local government, for example, still act as health and safety instructors, and a review of recent advertisements suggests that this is what is expected. In one organisation, 55% of the financial resource base was allocated to an activity occupying only 14% of the accident casualty base yet required 64% of the staff time to deal with it. In this particular case, objectives were not set, accident data were collected but not used and the safety unit carried out their duties in the same way that they had always done.

This may have been 'acceptable' practice in 1974, but it is now time to change. A company must operate within acceptable cost-benefit parameters. Financial restrictions now dictate a much tighter management approach to decision making and safety personnel now find that they have to be even more accountable than in the past. If they cannot justify their existence, then their relative priority within the company is diminished and provision will be reduced, as has happened in some companies. Regrettably, this poor management and presentation compounds the failure of senior management fully to recognise and comprehend the actual and the potential role of those required to carry out health and safety policy and practice, to recruit the right staff and train them properly and to provide them with sufficient resources to do the job. The organisational structure may be defective. Safety is no longer the 'Cinderella' area it once was. On the contrary, it should form an integral part of the strategic plan for accident reduction and should be given appropriate priority within the organisation and every office manager's duty to ensure the health and safety of their staff.

Those who are responsible for health and safety must improve the information on which decisions are based. This means that they should allocate sufficient resources to meet this requirement. Having spent time improving the quality and reliability of this primary data, consideration can then be given to its analysis for decision making purposes. From this, tactical objectives can be prepared for operational implementation and monitoring. Within this phase will be the determination of the evaluation criteria and performance indicators to be used. One of the important factors within the objective setting sequence is to quantify what cannot be achieved and to quantify more accurately why goals cannot be reached. You should develop a critical management style that provides for the identification and rectification of shortfalls. Board members, senior managers and members of safety committees should demand this information and it is right that they should be aware of all the relevant issues and not just told about what has been done. Remember to put the data into context.

While you might consider the implementation and monitoring of the company's safety policy there are other specific tasks you would be responsible for. These are summarised in Figure 1.3.

YOUR ROLE AS OFFICE MANAGER | 11

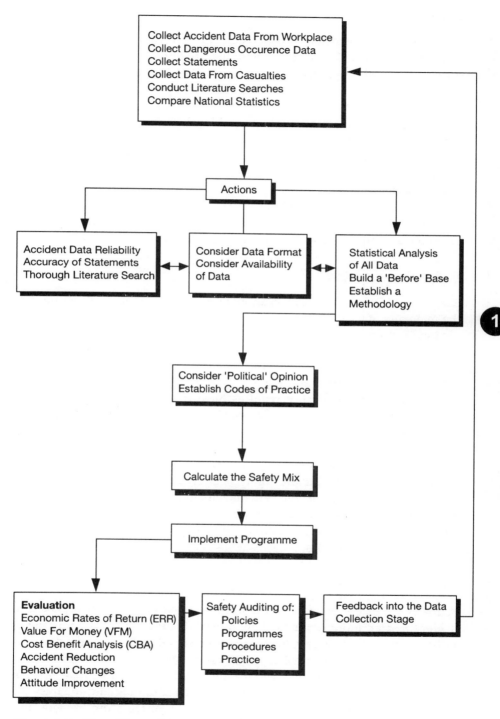

Figure 1.3 The office safety management process

The safety mix

The main aim of the manager responsible for health and safety is to co-ordinate the safety mix. This forms a main feature of the remedial process and successful safety programmes are based upon a mix of some or all of the main ingredients shown in Figure 1.4. It is vary rare for one part of the mix to be successful in isolation. Your task is to develop an appropriate blend within the safety mix framework at local level. This aspect is usually dealt with in the initial office safety programme planning stages.

Care should always be taken not to rely solely upon the enforcement aspect of the mix as the only means of providing an effective accident prevention strategy.

It is necessary that all these tasks are carried out efficiently and effectively and within clearly defined cost restraints. It is far better for the office manager to tell the accountants what is to be done rather than the other way round.

Safety monitoring

Safety monitoring concerns an examination of procedures and programmes, the main aim of which should be to help decrease the risk of accidents happening, and increase productivity. A systematic approach to office safety management will help you set objectives which will be based on priorities designed to improve management techniques, training and development, behaviour and employee attitude.

Such management practice requires measurement and this will need to cater for behaviour and attitude. Measuring these is not too difficult providing both are treated positively. Positive action will produce an activity which is measurable. A good attitude or behaviour must be developed and nurtured by management. This does not occur automatically. A genuine concern for the employee's health and safety is shown not only by management's position, but also by its action. Is management actively involved in accident prevention, or is it seen as a delegated interest? Employees are constantly evaluating the behaviour and attitude of management and this must be seen positively from both sides.

The safety review

A review of safety is geared to determine what action must be taken to achieve a safer, more productive operation. Its function is the gathering of information, but also is of value to management education and the development of an effective and efficient safety programme.

Many safety managers prefer to use the word *review*, rather than *audit*, in order, they say, to avoid the connotation of fault finding and to stress the more positive aspects of an investigative approach. The word 'review' also separates it from any similarity to a financial audit. Chapter 9 deals with the distinct difference between a *safety audit* and the *safety review*.

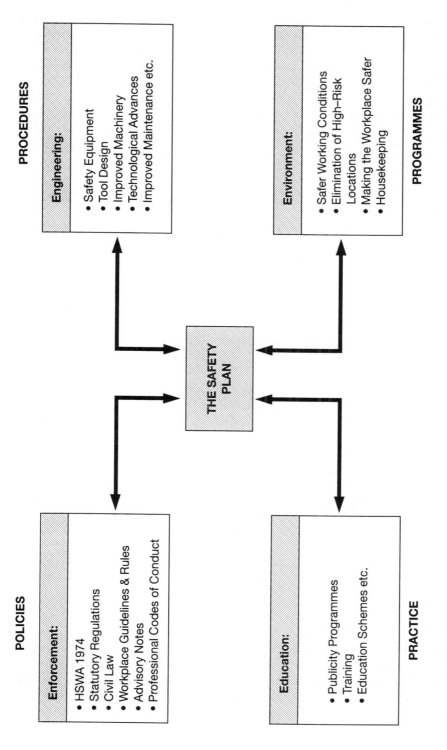

Figure 1.4 The safety mix

Safety reviews form a significant part of established management control functions. Via a system of interviews, the attitudes of employees, supervisors and management can be discovered. Any interviews with management should be held with the most senior executive possible. Although he or she will not be able to provide detailed answers to all questions, it is necessary to prepare well for the exercise. The reviewer or review team must provide an agenda to management in order that information can be gathered and people made available for the review.

There are advantages in obtaining external help when undertaking the first review:

- consultants are more objective in respect to the approach, design and concluding phase than someone within the organisation
- consultants aren't biased or influenced by traditions
- consultants aren't influenced by personalities
- independence can be very effective when changes are recommended.

The size and membership of a safety review team will largely depend on the complexity of the facility involved and the expertise that is required. It would be necessary to have periodic reviews every two to three years by an external expert, particularly if there are many changes in the organisation or operation.

Methodology

It is important to discuss the practicalities of the review process required to assist group executives and location management in the development or evaluation of their accident prevention activities. Although the areas for consideration listed below are summarised, they do outline the general topics required for discussion with senior management. A comprehensive safety review requires that it be conducted by qualified safety practitioners who will draw from their knowledge and experience so that useful information can be obtained. There are generally five review elements to consider:

- administration
- accident investigation and analysis processes
- job analysis
- self inspection
- contingency planning.

In addition to these, which are discussed further below, a number (the larger the sample the better will be the results) of supervisors and employees should be picked at random and interviewed to assess their safety attitude and to obtain data concerning their perception of the safety programme. It is important to select these people by occupation or trade rather than by name. This then removes, as far as possible, any prejudices for or against the management or its programmes.

A review should start with a conference with senior management in order to explain the agenda and to answer any questions that may arise. The exercise should end with an exit interview so that any recommendations can be fully explained and any differences resolved prior to the final report being published. Where an internal review is conducted, the team should obtain a conclusive response from senior management for each recommendation made. The scope of a rigorous safety review should be to evaluate the above five distinct elements in order to obtain answers to what is discussed below.

Administration

The interest of management and their full participation are the most important factors in any safety review activity. It is important to maintain this situation and secure management involvement. Office managers must be visible in the workplace for the promotion of safety as well as for production. In this way, it is possible to unite these two concepts to become safe production.

In a complex industrial or mining operation, there are often several facilities tied to an administrative or headquarters-type location. Individual sites sometimes have little or no control over areas of programmes or systems governing their operations, particularly those set up for the whole organisation. Examples might include the safety programme outline, the disciplinary programme or the needs of the workforce. It is important, therefore, that locations are evaluated for those systems which they control and that headquarters-type centres are analyzed for those that they command in the following ways.

Example safety review

HEADQUARTERS OR CENTRAL OFFICE

- Describe the loss control programme for:
  - employees
  - processes
  - contractors
  - equipment
- What is the office manager's role in the accident prevention programme?
- Is there a safety manager and to whom is he responsible?
- How often is the accident prevention programme reviewed internally for effectiveness?
- Is there a programme for training all office managers/supervisors based on established need?
- Does senior management occasionally attend safety meetings?
- Is there a system which recognises supervisors' and employees' safety achievement?
- Are aims and objectives required for safety performance?
- Are objectives realistic and measurable?
- Are objectives measured?
- What consideration is given to matching man, machine, and the environment in the hiring and placement of employees?
- Outline the programme for the selection and purchase of:
  - facilities
  - equipment processes
  - protective clothing and equipment

Site or satellite office locations

- Outline the accident prevention programme for:
  - employees
  - processes
  - contractors
  - equipment
- What is the role of management in the safety programme?
- Does management recognise supervisors and employees for their safety achievement?
- Does management initiate procedural and motivational communications to staff?

- Does management participate in accident prevention inspections of the facility?
- Does senior management attend occasional safety meetings with the staff?
- Are safety aims and objectives set jointly by the entire management team at that location?
- Are the objectives measured?
- Does senior management review all major incidents and subsequent statistical analyses?
- Is there a preventive maintenance programme?
- Is there a system of selection and placement of employees?
- Is there a job analysis programme in operation?
- Outline the safety training programme for management development and supervisory safety task training.
- Is there a safety co-ordinator and what are his responsibilities in relation to your office?

Accident and dangerous occurrence investigation

- Are all incidents, including first-aid, investigated for type and contributory factors?
- Do remedial measures attack the factors contributing to the incident happening?
- What system is used to follow up on preventive measures?
- Does the investigation indicate the potential for severity?
- Are all fatalities, major injuries and illnesses, and high cost incidents investigated?
- Does each department head review all incidents in his/her department?
- Does the safety department review and follow up on all incident reports?
- Are thorough analyses made of all incidents for trends? Are these analyses reviewed by senior management?
- Is there a method of determining the number and type of incidents resulting in lost time from off-the-job injuries and illnesses?
- Are off-the-job incidents analysed for type and costs to the facility? Are these analyses communicated to senior management?
- Do staff safety awareness programmes include off-the-job safety?
- Is off-the-job safety information provided to the families of office staff?

Job analysis

- Has a priority list of hazardous or high risk jobs been established?
- Are hazardous/high risk jobs analyzed by method to determine the potential for accident and the safe work procedure?
- Are the priority listed jobs necessary to accomplish the work aims?
- Who participates in the job analysis?
- Is job or task training based on the completed job analysis?
- Are safe work procedures developed through job analyses?
- Are they used as guides:
  - by staff to perform work?
  - by office supervisors to train and supervise?
  - by office managers to manage?
- Are the job analyses reviewed when:
  - incidents occur?
  - there is a change in equipment?
  - there is a change in environment?
  - there is a change in the process?
- Are equipment pre-use or inspection check sheets developed using the job analysis procedure?
- Are tags and signs developed and used to support the safe work procedures?

Self inspection

- Is there an inspection team with representation from management, production, safety and office staff?
- Does the team have a checklist including all areas of the facility?
- Is there a system of follow up for the remedy of unsafe conditions and practices observed?
- Does the inspection include the preventive maintenance programme records?

In addition to the overall plant inspection, are periodic inspections made:
- by department heads and supervisors of individual areas?
- by a safety practitioner?
- by senior management?
- by the office manager/supervisor?

Contingency planning

- How many emergency medically qualified personnel are there on staff available at any one time?

- How many are trained in first aid?
- Where are the first-aid stations located?
- How often are the first-aid stations checked for operational efficiency?
- Is there a doctor on retainer or designated?
- Are there clear written emergency and communication procedures?
- Are the responsibilities of management and supervisors clearly defined in these procedures?
- What is the availability of ambulance services if required?
- Who administers or co-ordinates the emergency procedures?

A safety review places the emphasis and encourages the activity where they will produce the best results. Office management and supervision are key factors in effective safety programmes. It is the responsibility of the office manager to ensure that such reviews are held regularly and that the whole workforce including senior management are involved. It is positive action, rather than what is said, which is of vital importance in demonstrating the management position.

Table 1.1 Your health and safety checklist

Task	Response
Who is in charge of the job?	
Has anyone else a responsibility for the job?	
Can you identify any task which nobody is responsible for?	
Are there established safe ways of doing the job?	
What are the guidance notes or codes of practice?	
Have you provided safe working procedures to do the job?	
Do you need to make the job safe?	
Do you need protective clothing?	
Do workers know the limitation of safety equipment and clothing?	
Has anyone assessed the equipment to be used in terms of suitability?	
What will be the consequences if it goes wrong?	
Can emergency services get to the site?	
If the job cannot be finished today can it be left in a safe state?	
Are there clear enough instructions for the next shift?	
Does everyone else know what is going on?	
Has the job been done in the way intended?	
Has anyone checked it?	

ACTION CHECKLIST

- Do you now understand your safety role within the organisation?
- Do you understand what the requirements of RIDDOR are?
- Do you keep, maintain and check your accident records regularly?
- Do you know what a Form F2508 and F2508A are for?
- Do you use accident data for decision making purposes?
- Do you use supplementary information to enhance your accident data base?
- Who do you involve in safety management decision making?
- How are your health and safety programmes implemented, monitored and evaluated?
- Do you understand the safety mix principle?
- How often do you evaluate your safety policies, programmes, procedures and practices?
- Is your accident situation better or worse than last year?

❝You must ensure that appropriate safety systems are in force and that they comply with the law❞

❝Failure to comply will ... result in the prosecution of your organisation and in certain circumstances you...❞

Legal requirements

INTRODUCTION

It is impossible to discuss office health and safety without spending some time outlining the relevant law on the subject. The office is usually the first port of call for a visitor and it is essential that the visitor does not get the wrong impression of your organisation right at the outset. The author once visited a well-known mining company and on entering the main office was immediately entangled by a printer cable which had been affixed to the top of the door frame albeit very loosely.

As office manager with a responsibility for health and safety you must ensure that appropriate safety systems are in force and that they comply with the law. You must also ensure that statutory regulations are obeyed and that both the general policy and strategic decisions relating to safety are in place. It is essential that these are translated on a daily basis into detailed plans and procedures for staff to follow. Finally, you must insist that these systems operate effectively without any lapses. Generally the law requires the following:

- safe systems exist within the office environment
- effective maintenance of both buildings and equipment within it
- staff receive appropriate and adequate:
  - safety training
  - information about health and safety procedures and practice
  - instruction about safe practices and policies
  - supervision when undertaking hazardous or dangerous activities
- safety of others including the general public who may be affected by operations of the organisation
- third parties such as contractors or consultants hired by the organisation do not endanger staff nor are endangered by them.

Failure to comply will almost certainly result in the prosecution of your organisation and in certain circumstances you can be held *personally* liable for both criminal and civil proceedings.

This chapter only summarises a rather complex subject and health and safety law is more comprehensively covered in other publications. In this chapter, therefore, it is intended only to trace the developments in certain legal areas which are of interest to office managers. For more serious students of the law a reading list is given at the end of the book.

Throughout the text, any reference to safety will include all health issues and any other areas which are hazardous or dangerous.

Enforcement strategies operating within the legal framework do, however, impinge on those responsible for safety, particularly when things go wrong. It is your responsibility to know when plans under discussion need expert advice and to seek it from qualified legal staff or appropriate technical consultancies. You will need a system of being kept up to date with legislative improvements and changes. Membership of one or more of those organisations listed in the Appendix is highly recommended.

When examining safety-related case law, it can be concluded that it has been established as the result of the safety management system failing to meet the aims and objectives of the organisation. Although companies do not maliciously set out to injure their workforce, there are many examples where 'other factors', usually of a financial nature, have compromised the need for an effective safety policy. In such instances, it would seem that safety is regarded as a worthwhile priority by an organisation until a conflict arises between safety and the balance sheet. Safety management must be seen in *profit* rather than *expense* terms. To illustrate these points it is helpful to know some of the relevant issues regarding safety and the law from an office management standpoint.

HISTORICAL BACKGROUND

Common law is a part of our legal system which is regarded to be traditional and is effective in areas not covered by Act of Parliament. The principles and rules of common law are contained in those decisions taken previously in English courts. They are recorded in the law reports and go back to the Middle Ages. Health and safety issues began to appear in case law during the industrial revolution when casualties began to sue for compensation following injuries received at work. Common law is based on what has been decided before by judges and is referred to as precedent. Although there are a few technical exceptions, a court is bound to follow those decisions made earlier in courts of equal or higher status, unless statute law dictates to the contrary. Common law governs the rights and duties of individuals towards one another.

Generally there are four remedies that are available to an individual:

- *compensation* – a cash value is placed upon the injury suffered or the loss experienced where an employer is found to be liable through failure to comply with the law
- *reparation* – where an employer could be required to restore conditions so that they are the same before the breach of the law
- *performance* – where an employer can be compelled to perform his obligations

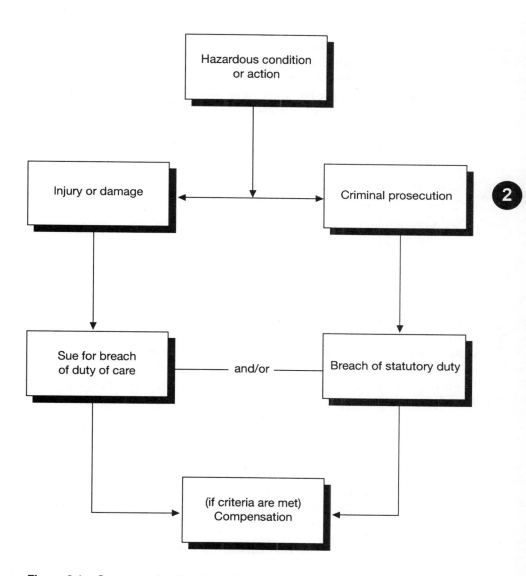

Figure 2.1   Courses of action in a claim for compensation

- *injunction* – where an employer can be required to desist from an activity where it interferes with the common law rights of another.

Civil actions can be initiated in a variety of courts such as the County Court. It is said that common law can be traced back a thousand years although references to safety matters began appearing in the mid-nineteenth century. This coincided with the industrial revolution and the advent of the factory type of workplace.

Under common law, workers could sue their employers in the courts for compensation for injuries received whilst at work. Common law requires that an employer must take reasonable care to protect his employees from risk of foreseeable injury, disease or death at work. In the nineteenth and early part of the twentieth century, employers argued with reasonable degrees of success against this duty in respect of proving the 'foreseeable' aspect of the law. It was not until 1938 in *Wilsons & Clyde Coal Co Ltd v. English* ([1938] AC57) that the House of Lords identified in general terms, the duties of an employer at common law. They judged that all employers are required to provide and maintain:

- a safe place of work
- a competent staff of men (or women)
- a safe system of work
- safe plant and appliances.

Also, employers were made liable for injury causing accidents to their employees as a consequence of the negligence of other of their employees, provided that the act of negligence arose out of and in the course of employment. This principle is the effect of the application of *vicarious liability* at common law. This is a similar standard to which an employer must conform under the Health and Safety at Work Act which came into being in 1974, and is discussed below.

Employers have a *duty of care* and here, Lord Atkin established two basic principles:

- you must take reasonable care to avoid acts or omissions which you can reasonably foresee would be likely to injure your neighbour.

Who then, in law, is my neighbour?

- persons who are so closely and directly affected by my act that I ought reasonably to have them in contemplation as being so affected when I am directing my mind to the acts or omissions which are called in question.

Where accepted practice can normally be used in defence, a duty of care extends to the employer considering whether the practice is adequate and the need to change it if necessary. Some cases highlight a reactive management strategy whereby no action is taken until the law provides for it. Such management strategies are open to action of this nature, particularly if an organisation chooses to allow courts of law to make their safety management decisions. An over enthusiastic safety policy can close most factories by making them too safe. In this instance, it can be said that a workplace can only be deemed to be safe when it is shut! Risks have to be accepted, therefore it is essential that safety management is proactive by nature. Risk analysis and an historical accident and dangerous occurrence data base are prerequisites. A court is influenced if safety management decisions are made for sound reasons. Management strategies that are incompetent are easily shown to be negligent by courts.

EMPLOYER LIABILITY

Whilst the requirements of the law have been briefly discussed there is one aspect of the enforcement strategy which is important. This is in the area of discipline. Many office managers need to appreciate that the enforcement strategy includes the safety rules which operate within an organisation. Breaches of discipline must be dealt with quickly and fairly. If an employee is in breach of a rule or rules than a fair system for dealing with this must be established. The ultimate disciplinary action here would be dismissal.

In law, the duty of care may require an employer to discipline or even dismiss an employee who is a source of danger to his work mates. This was proven in *Hudson v. Ridge Manufacturing ([1957]) 2 QB 348)* where an employee continually took part in dangerous horseplay against his work mates. Despite several reprimands from his foreman to desist a worker was eventually injured as a result of the skylarking. The company was held liable as reprimands are not enough. Another implication of this case is discussed subsequently.

Reasonable care and the breach of duty

The test of reasonable care is determined by a court on the basis of what a reasonable man would do in the same circumstances as a defendant. If the existence of the duty of care is established as a matter of law then the breach of this duty is decided by the judge.

There are five factors in determining reasonable care:

- *cost* – the amount of money that it is necessary to spend against a slight possibility of risk is limited
- *obviousness of risk* – the more obvious the danger the more liable the employer will be held for failing to prevent an accident. A partial defence to this is that the employee may have been aware of the risk, in which case the employee can be sued for contributory negligence
- *inherent risk* – all work carries with it some possibility of risk which is irreducible or irremovable for which the employer cannot be held responsible
- *likelihood of injury* – the greater the risk, the greater the liability

- *seriousness of injury* – the more serious the consequences, the more precautions should have been taken

Burden of proof

'He who asserts must prove.' Generally the plaintiff must prove that the action or omission of the defendant actually caused the injury. However the rule of *res ipsa loquitur* (the thing speaks for itself) assumes that the only possible explanation for the injury was the defendant's negligence. In this case the onus is on the defendant to prove that this was not the case.

The plaintiff is required to demonstrate cause and effect. The plaintiff must show the defendant's negligent act or omission caused the injury or loss:

- a defendant is only liable for the damage which a reasonable person would have foreseen as a likely consequence of an act
- a defendant is only liable to compensate for damage that was foreseeable and not all the direct consequences of the act.

Where the consequences of an act are direct, where one consequence flows automatically from another and cannot be avoided, than the *Re Polemis* rules will apply. However, if there is an interruption, diversion or an intervening and new physical event, the chain of causation is broken and the result will be held to be too remote.

Vicarious liability

This is defined as where one person is held to be liable for the behaviour of people acting on his or her behalf. It would be unfair to make an employer a target for a common law claim as a result of all negligent behaviour so a set of rules has been established by which to judge the extent of this liability. An employer may be liable for the torts of his employees and those of independent contractors (a tort is a civil wrong).

Liability for the torts of employees exists where:

- a wrongful act or omission has been expressly or impliedly authorised by the employer
- a wrongful act or omission is committed by doing something authorised by the employer in an unauthorised manner
- a wrongful act or omission is committed and ratified by the employer.

Before liability can be proved it has to be shown that:

- the employee is a true employee (is paid and is subject to normal procedures of hiring and dismissal)
- the employee has committed a tort or crime
- the tort was committed in the course and scope of his/her employment, for example doing what he/she was employed to do.

Specific defences

There are several defences available to an employer which either rebut the allegation of negligence, or attribute the damage or accident to a circumstance which is beyond their control. He or she can plead that as they took reasonable care there was no negligence or that no duty of care was owed. There are also some specific defences.

Novus actus interveniens
'A new act has intervened.' This covers the actions of a third party between the negligent act and the subsequent damage to the injured person. It is the function of the court to decide if the intervening act could have been reasonably foreseen by the defendant.

Volenti non fit injuria
'No harm can be done to a willing person.' This defence is used to show that an injured person was aware of the risk involved and consented to it.

Contributory negligence
This is used to describe the behaviour of an injured person whose actions contributed to the occurrence giving rise to his or her injuries. Until the Law Reform (Contributory Negligence) Act 1945 this was held to be a complete defence. Now damages shall be reduced for the plaintiff according to how much their own negligence contributed to their injury.

Agony of the moment
This is a defence against contributory negligence.

Common law implications

An employer's duty consists of the provision of safe plant and materials, competent staff and safe systems. **Therefore as an office manager responsible for safety you must make sure that the:**

- employee is aware of any dangers associated with the job
- employee knows what precautions must be taken in view of those dangers
- precautions that are necessary are always available for use
- employee knows that the precautions are available for use.

To achieve these aims an energetic safety campaign has to be undertaken to get the message over to employees. You will not have discharged your common law duty of care simply by issuing safety equipment or protective clothing if you do not educate your employees as to why these are necessary. You should be able to demonstrate that you have undertaken group and/or individual training exercises. Further, that you ensure so far as it is reasonably practicable, that employees take these precautions. Supervision must ensure that precautions made available by the employer are used, otherwise the employer can be held liable in the event of an accident.

Non employees
In essence the degree of responsibility will be proportional to the amount of control the occupier has over the visitor. Sometimes the occupier may have so much control over the activities of the visitor or user that their duty is as high as if he or she were employer and employee. This is particularly true where the visitor is a contractor's workman working under the instruction of the occupier. The civil liability of an occupier of land or buildings or fixed or movable structures such as vehicles and scaffolding towards persons coming lawfully onto those premises is laid down in the Occupiers Liability Act 1957.

The extent of an organisation's responsibility will depend on the type of work being carried out. If one employment is more dangerous than another, a greater degree of care must be taken by the employer. If employers are not able to eliminate the risk, they must at least take reasonable care to reduce it as far as possible. Employers must take reasonable care to ensure the safety of their employees' workplace even when it involves work at another occupier's premises. However in the latter case, the employer can only be expected to take limited precautions. The employer must ensure, however, that proper safety equipment and instruction is given.

Under section 2 of the Occupiers Liability Act 1957, the occupier must take reasonable care to see that their visitors are reasonably safe in using the premises for the agreed purposes of the visit. The same duty is owed to all lawful visitors, including people with a legal right of entry such as HSE inspectors, utility engineers, etc. The Act makes provision for the circumstances of the visitor, for example, children will be less careful than adults.

The organisation has a twofold duty of care with regard to contractors, both a responsibility to them and for them. Liability rests with the person in control (the occupier). Control, however, may be divided between different occupiers.

However, there are two circumstances which are likely to put liability back onto your organisation. The first is where your own employees have recourse to the area occupied by the contractor, (you are still, in effect, occupying that part of the premises and have a personal duty to your employees). Secondly, where your organisation has been negligent in the choice of contractor and has selected someone who is not competent to do the job safely. If the dangers are caused by contractors working on the occupiers' premises, the Act states that the occupier is not liable if he has taken reasonable care to see that the premises are safe.

There are three circumstances where the manager and the organisation cannot escape liability:

- where the employer asks the contractor to break the law
- where the contractor is employed to undertake extra-hazardous activities

> **Checklist**
>
> - Have they done this kind of work before?
> - How much experience do they have?
> - Can anyone furnish references as to the quality of their workmanship?
> - What provisions will they make for the safety of my employees?
> - What guarantees do they give for completion and performance of contract?
> - Have you specified safe working practices?
> - Have you checked compliance on a regular basis so far as it is reasonably practicable?

'Any person who brings onto his lands anything likely to do mischief and it escapes, must keep it at his peril, or be answerable for the consequences should it escape'

- liability to visiting contractors. Essentially occupiers are liable to other people's workers for risks which are not occupational hazards, but which are peculiar to premises and of which they themselves know or ought to know.

CRIMINAL LIABILITY

Acts of Parliament impose duties upon all of us. For example, we must insure our vehicles if we drive one on the public highway, sell goods fit for the purpose, not allow certain rooms to fall below a certain temperature, etc. These duties are enforceable by punishments which are usually decided by a court unless a fixed penalty has already been agreed and established. Statute law governs matters that affect the state as a community. In safety management terms, certain statutory laws allow for a person injured as a result of a breach of this law to bring civil proceedings against the other for the injuries received. However, most, if not all statutes, place general standards of conduct upon people and these are usually not very specific. It is then up to judges to interpret the law in order to test whether such action was meant and covered by the statutory act.

This may also mean that an employee injured as the result of a breach of statutory duty may also claim as a result of negligence. It is equally feasible to win on one point and lose on the other as it is to lose both or win both. It must be said that an employee winning on both does not necessarily get double damages.

Failure to comply with a statutory duty is a crime and prosecutions can be initiated in courts such as the Magistrates' Court or County Court.

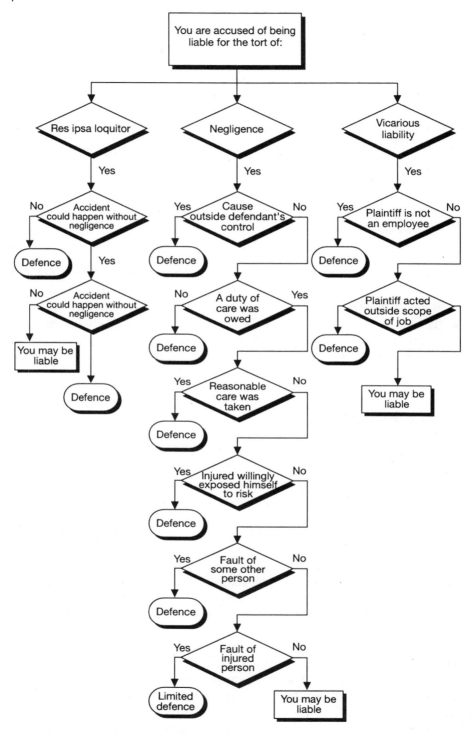

Figure 2.2 What the office manager should know

For example, a statutory duty is placed on the occupier of a factory under the Factories Act 1961. Generally the issues contained in the Act are:

- health
- safety
- welfare.

The first covers such things as cleanliness, ventilation, toilet facilities, lighting, working temperatures, drainage and overcrowding. The second concerns machinery fencing, hoists, chains, obstructions, fire escapes, etc., whilst the third provides for first-aid facilities, washing and rest facilities, provision of drinking water, etc. These points are quite specific and action for damages can be brought for loss suffered through a breach of these duties and responsibilities. The Factories Act 1961 applies to all factories of whatever size, irrespective of the number of employees and the type of trade or business being carried out. The definition of a factory is very long and appears to be based upon the employment of manual labour. An example to illustrate this is in respect of the canteen. A canteen within the physical bounds of the factory used for feeding and entertaining the workers is deemed to be part of the factory yet a restaurant used solely by management is not. This is based upon the distinction that feeding and entertaining the workers is not incidental to the normal processes of the factory, but to feed the management is. A person injured whilst cleaning the workers' canteen would get damages whilst a person injured whilst cleaning the managers' restaurant would not!

Whilst the factory is covered by the Factories Act 1961, the Offices, Shops and Railway Premises Act 1963 covers the occupiers of offices, shops and railway premises. The Act deals with the same issues covered by the Factories Act and includes advice on health, safety and welfare issues. Workers in agriculture and forestry were similarly provided for in the Agriculture (Safety, Health and Welfare Provisions) Act 1956, whilst the Mines and Quarries Act 1954 provided rules for the control and management of mines and safety including roof supports, ventilation, protection against dust, winding apparatus, construction and fencing of machinery and the use of explosives. Under all three Acts, an action for breach of statutory duty lies on the same principles as apply to the Factories Act.

At this time, there were still some employees who did not enjoy statutory protection. It was not until the introduction of the HSWA that full protection was provided. It gives protection to all at work (except those in private domestic employment) including those workers on oil rigs and other offshore installations. Earlier laws were unaffected by this piece of legislation and the HSWA enables regulations and codes of practice to be altered with relative ease. This is a particularly important feature since technology in some industries is known to move faster than the law. An example of this occurred following the Flixborough Nypro Chemical explosion in 1974. The

Health and Safety Commission established an advisory committee on major hazards and it published new guidelines in 1976.

The main thrust of the HSWA comes in section 2 where an employer's duties to employees are discussed:

- the provision and maintenance of plant and systems of work that are, so far as is reasonably practicable, safe and without risk to health
- arrangements for ensuring, so far as is reasonably practicable, safety and absence of risks to health in connection with the use, handling, storage and transport of articles and substances
- the provision of such information, instruction, training and supervision as is necessary to ensure, so far as is reasonably practicable, the health and safety at work of employees
- so far as is reasonably practicable as regards any place of work under the employer's control, the maintenance of it in a condition that is safe and without risks to health and the provision and maintenance of means of access to and egress from it that are safe and without such risks
- the provision and maintenance of a working environment for employees that is, so far as is reasonably practicable, safe, without risks to health and adequate as regards facilities and arrangements for their welfare at work.

Reasonably practicable

Generally, professional safety practitioners find the legislation of immense value but are frustrated as to the vagueness of the repeated statement 'where reasonably practicable'. What might be deemed by one to be reasonably practicable, may not be deemed so by another. This confusion can only be made clearer by examining how judges have interpreted this matter in court.

General duties of employers

'Reasonably practicable' is a narrower term than 'physically possible' and seems to imply that a computation must be made by the (employer) in which the quantum of risk is placed on one scale and the sacrifice involved in the measures necessary for averting the risk (whether in money, time or trouble) is placed on the other; and that, if it is shown that there is gross disproportion between them – the risk being insignificant in relation to the sacrifice – the defendants discharge the onus on them.

Onus of proof

Unlike the common law which places the burden of proof on the plaintiff, the HSWA places the onus of proving the limits of what was reasonably practicable with the accused.

It is important to show the court under these circumstances that this assessment was made before the charges were laid, for example before the alleged offence had taken place. Lack of money or resources is no defence.

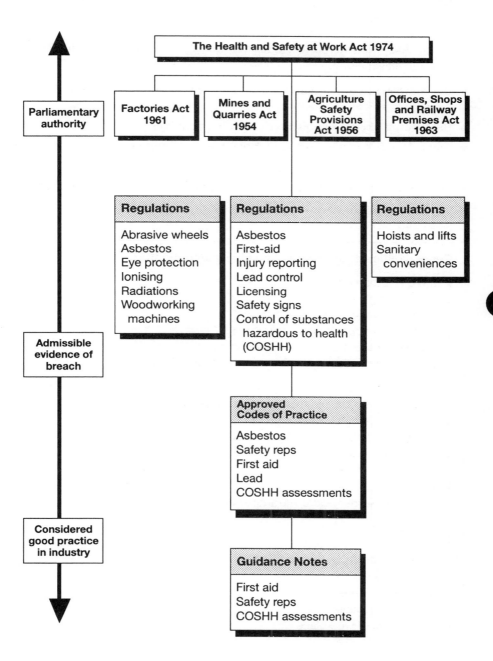

Figure 2.3 The HSWA 1974 and its relationship to other statutes and regulations

Section 2 of the HSWA goes on to outline the general areas to be considered by the employer when complying with the general duty previously described.

Factors to be considered when discharging your duty towards contractors and employees include:

- the competence or specialist skills possessed by the contractors and their workforce
- the amount of control you have over their activities
- whether or not you are providing plant and equipment for their use
- workplace dangers that the contractor ought to be warned of
- areas which may contain potential hazards of which you are not aware.

General duties of employers – employees/contractors

- fire precautions taken
- sufficient lighting
- sufficient working space
- clear unobstructed gangways
- tidy workplace, floors in good condition
- steps, stairs, ladders, scaffolding etc. in good condition

- dangerous machinery to be guarded
- regular inspection and maintenance procedures
- emergency stop buttons
- sufficient equipment of the right quality
- correct procedures laid down
- safety equipment
- clear instructions
- permits to work in extra-hazardous areas

- adequate ventilation
- reasonable temperature
- accidents promptly investigated
- first-aid facilities
- washing facilities
- sanitary conveniences
- cleanliness of workplace
- waste disposal, etc.

- instructions, training, etc.
- protective clothing and equipment mechanical handling equipment
- forklift trucks, conveyors, trolleys, etc.
- storage areas defined

General duties of employers – physical aspects

- safe place of work with a safe means of access and exit
- safe plant and equipment – including maintenance
- safe systems of work
- safe working environment and adequate facilities and arrangements for employees' welfare
- safe methods for handling, storing and transporting goods.

Duties concerning the physical aspects of the work environment and of the people who are employed:

- provision of information, instruction, training and supervision
- trade unions must be allowed to appoint safety representatives from within their membership. Management has no say in these appointments
- employers must prepare and revise, where appropriate, a written statement of policy in respect of health and safety (safety policy). This applies where an employer employs more than five people.

Information, instruction and training

The provision of information, instruction, training and supervision ties in very closely with your common law duty to provide competent staff. Not only so that the interests of fellow employees are guarded but also so that the individual employee is better equipped to take care for their own safety.

These provisions include such things as fire extinguishers, first-aid equipment and facilities, protective clothing and equipment, guards on machinery and so on.

The office manager now has an opportunity to compare the responsibilities and those duties at common law.

It is important to recognise the extent of the employer's duties in relation to those of the employee. It would seem that the employee will only be liable if he/she does something outside the control of his/her employer and/or unknown to him/her, or in direct contravention of instructions given by his/her employer.

The amount of reasonable care or co-operation to be expected of the employee is going to depend heavily on the amount of instruction, training and information he/she has been given. The amount of supervision necessary will equally depend upon these three things.

As the amount of information, instruction and training given to the employee increases the amount of supervision will decrease. This does not mean to say that supervision is no longer required but that it may be reduced to take account of the skill and experience of the individual employee.

Law of statute to non-employees

Employers also have a duty under statute to people who are non-employees. This involves the actual operation of the business and working practices. Concerning responsibilities involving the premises and ownership of plant and materials, the relationship between the employer (as a client) and independent contractors has been the subject of much debate. Where the organisation's employees are called to work jointly with the employees of contractors, one of the questions which might arise is that of responsibility for the provision of information, instruction and training for the two groups of employees.

It is important for employers to ensure that the provision of information about dangers peculiar to their places of work or processes extends to persons other than their own employees who may be affected by them or whose actions may affect the health and safety of their own employees.

In addition it would be advisable to provide instruction on the safe use of equipment available on the premises or on the proper precautions to be observed to minimise the risk of accidents occurring. It may also be necessary to give verbal or written instruction, to provide safety manuals or information leaflets and, in special circumstances, to provide training to persons other than direct employees.

These responsibilities would include the following typical areas:

- ensure staff are trained and competent to carry out their tasks
- ensure sufficient information and instruction is given
- ensure plant, machinery and equipment in your department is safe and properly maintained
- ensure safe access and egress within your department
- ensure a safe place of work
- ensure fire precautions are taken.

Hybrid offences

All other offences specified by the HSWA can be tried in court either way, for example:

- failure to discharge a duty imposed by sections 2–7
- contravention of sections 8 and 9
- contravention of any health and safety regulation
- contravention of an improvement of prohibition notice
- illegal disclosure of certain kinds of information
- making a false statement or entry to a register or document, or forge a document with intent to deceive
- impersonating a factory inspector.

Only certain offences actually carry the full sanction of a prison sentence and these are as follows:

- a breach of the requirements of a prohibition notice
- a breach of the conditions of issue of a licence
- attempting to acquire explosives
- the disclosure of information given under certain circumstances by the HSC.

The first ever prison sentence as a result of a prosecution under the HSWA was given in 1985 to a company director who 'demonstrated a cynical disregard for the dangers to health'.

The individual was also personally fined a total of £1,500 along with fines to his company totalling £2,000. The director failed to comply with a prohibition notice served on the removal of asbestos lagging. His one-month prison sentence was suspended for two years.

Trades unions see this training as purely their province and representatives are entitled by law to attend a TUC approved course. This principle has been tested in an industrial tribunal. However, this should not be the beginning and end of their training. Union courses should be supplemented by in-house training dealing with specific issues involved in the organisation work. Ideally the training should produce a balance between all three roles previously mentioned.

Further guidance

Other means of guidance and advisory literature will generally fall into one of the following categories:

- *guidance notes by the HSC* – these will be authoritative in so far as they carry the Commission's backing but they will fall short of approval of any particular course of conduct by an employer
- *guidance notes by the HSE* – these will be essentially working notes of guidance. They will for example include advice on requirements to be followed and action to be taken by employers in order to comply with the law, technical data about specific hazards, and the interpretation of biological data by medical officers
- *guidance notes by industry advisory committees* – these will originate in the form of reports which will carry the authority of the individual committee
- industry-generated notes of guidance may come from manufacturers and other associations providing technical and common sense instructions on how to cope with various situations in particular workplaces. The HSC or the HSE will normally not wish to give any official recognition to these notes of guidance.

Failure to comply with the requirements may lead to prosecution for breach of a statutory duty. Most safety cases are heard in the Magistrates' Court summarily but the more serious ones are heard on indictment in the Crown

Court. Do not forget, unlike common law there does not have to be an injury before action can be taken by the HSE inspector (see below).

HEALTH AND SAFETY AT WORK ACT 1974 (HSWA)

The HSWA (the Act) is likely to impinge most directly on office managers who are held primarily responsible for the compliance with this Act in their company. The Act determines specific standards of safety in a number of areas through a number of regulations issued under its provisions. The Act imposes a number of legal duties on employers, the self-employed, employees, suppliers and owners of premises, to ensure that their workplaces are safe and offer no risk to health. The Act's four basic aims are to:

- secure the health, safety and welfare of persons at work
- protect persons, other than persons at work, against risk to health and safety arising out of or in connection with the activities of persons at work
- control the keeping and use of dangerous substances
- control the emission into the atmosphere of noxious and offensive substances

To achieve these aims the Act established one comprehensive, integrated system of law dealing with health and safety and welfare of people at work and the public who may be affected by such work activities, and in so doing:

- places general duties on the employer, ranging from providing and maintaining a safe place of work to consulting with employees
- places duties on employees to take reasonable care of themselves and anyone who may be affected by their work activities, and to co-operate with their employer and others in meeting these obligations.

The Act is a piece of enabling legislation which allows the Secretary of State to define a system of regulations and approved codes of practice which will progressively replace the existing legislation, but maintain or improve the standard of health, safety and welfare established by the Act. Thus the regulations can be amended to keep pace with developments.

Implications of the Act

Accountability

It can be seen that the Act in sections 2, 3 and 4 imposes wide ranging and far reaching duties on the organisation by virtue of its position as employer and controller of premises. However the Act does not stop with the inanimate 'body corporate'. Section 7 makes it the employee's duty to have a care for himself and for others who may be affected by his acts or omissions at work, also placing on him a duty to co-operate with the employer to enable the requirements of sections 2, 3 and 4 to be complied with. Sections 33, 36 and

37 make it an offence for an individual to fail to discharge the duties placed on him by sections 2–7 and remove the 'protection' afforded by the body corporate. Not only can the organisation be prosecuted for offences under the Act, so can the managing director and his subordinates right down the management chain, where it can be established in law that they have failed. Therefore, wherever a person holds a management position, the responsibility conferred with that post includes responsibility under the Act commensurate with the status of the post. Further to this, section 8 refers to 'persons' who shall not interfere with or misuse anything provided in the interest of health, safety and welfare. Thus any person who misuses or vandalises a first-aid box or fools with a fire extinguisher can be prosecuted under this section of the Act.

Control of premises
Section 4 of the Act is applicable in the situation where persons who are not employees of the organisation including members of the general public, are on the premises of the organisation. The plant or substances can range from an ordinary chair to an arc welder, or from water to a radioactive isotope.

Safety information and training
Section 2(2)(c) requires the employer to provide information, instruction, training and supervision. Much of the requirement devolves on the responsible office manager as follows:

- *information* – this will be passed from the department or section. Where applicable and relevant, this information must be brought to the notice of affected employees by the responsible manager. Additionally, information applicable to the safety of a particular process or piece of equipment must be readily available for employees to study

- *instruction and training* – the basis of much of health and safety is effective instruction and training. Whilst much of this will originate from and be organised by the responsible department or section, the manager has to ensure that all members of his staff receive the relevant and required instruction and training. Additionally, because of sections 3 and 4, in certain circumstances 'persons not employees' may also need to be trained

- *supervision* – the appropriate level of supervision must be exercised at all times. It is not sufficient to provide information, instruction and training without adequate supervision in safety matters. Such supervision must primarily be directed at ensuring that all subordinate safety staff are aware of their responsibilities. You must consider the adequacy of the level of supervision in possible hazardous areas, particularly in the light of the experience of the employees involved.

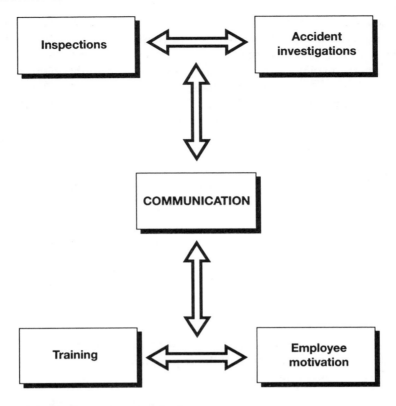

Figure 2.4 Performance appraisal

Duty not to charge
Section 9 places a duty on the employer not to charge the employee in respect of anything done or provided in pursuance of any specific requirement of the relevant statutory provisions. Thus, where there is a statute law in the form of a regulation requiring certain matters to be done, such as the Protection of Eyes Regulations which requires operators to have eye protection, this must be provided free of charge. Where practicable all protective clothing should be provided free of charge.

Duty of manufacturers
Section 6 places a duty on the manufacturer and the supplier broadly to make and supply articles and substances which are, as far as is reasonable practicable, safe and without risk to health for use at work. Employers and individual managers must ensure that instructions provided by a manufacturer or supplier are brought to the attention of and obeyed by employees.

Suppliers' duties (section 6)
One aspect of the HSWA not discussed so far is the responsibilities it places upon the supplier of equipment and materials. In fact section 6 covers the

responsibilities not only of suppliers but also of designers, manufacturers, importers and installers relating to articles and substances for use at work. These are to:

- ensure that these articles and substances are safe and without risk to health when properly used
- carry out tests or examinations as may be necessary to ensure that they are safe and without risk to health when properly used
- provide any information necessary to ensure that they are safe and without risk to health when properly used.

Installers of equipment are under a similar obligation to ensure that anything they install is safe and without risk to health. These duties can only be relieved by a written undertaking from the user that he/she will personally take steps to ensure that the article or substance will be made safe.

Initially you may feel that these duties have no relevance to you or your organisation or may even relieve you of your responsibility as a user of equipment. There are however two important factors which you must consider.

First as a user of equipment you have a responsibility to your employees to provide (so far as is reasonably practicable) equipment and materials that are safe and without risk to health. You cannot rely upon the integrity of the supplier in every instance to protect the safety of your staff. It is necessary for you, the user, to request information from the supplier, etc. to enable you to use the article or substance safely. Furthermore, if an article or a substance has a defect or is a risk to health and safety and it can be shown that you were (or should have been) aware of this, it is no defence to argue that it was supplied in that condition. You may share liability with the supplier.

Secondly, the organisation may itself be a supplier of articles or substances for use at work and consequently may incur the liabilities imposed by section 6. This often happens where items surplus to inventory are concerned and disposed of by selling them to others who may intend to use them at work.

Safety representatives and committees

A unionised employer has a duty to consult with safety representatives appointed by the appropriate trades union. Where two or more safety representatives require the formation of a safety committee, an employer must undertake the establishment of such a committee under section 2(7) of the HSWA and the Safety Representatives and Safety Committees Regulations 1977, reg. 9(1) within three months of the request being made.

ENFORCEMENT

Within the enforcement area of office safety management is that of policing. To ensure that organisations are conducting their business within the terms of the HSWA, the HSE was set up. At the same time, the Health and Safety Commission (HSC) was created to oversee the work of the HSE. The HSC has the general overseeing of the work of the HSE and has power to delegate to it. However, the HSE is a statutory body charged with the responsibility for the enforcement of safety legislation. The HSC cannot give instructions or directions to the HSE about enforcement matters and cannot issue judgments in any particular case. Both HSC and HSE are rather bureaucratic and cumbersome.

Safety laws within your premises are enforced by the HSE. Primarily this agency will seek to advise and assist employers in meeting the appropriate standards although it does have broad powers of enforcement.

The HSE inspectors have a number of powers available to them and when appointed they are issued a warrant card which specifies the range of these. These powers enable them:

- to enter premises at any reasonable time
- to take a constable if necessary
- to take any authorised person and equipment, for example, gas, electricity engineers
- to examine or investigate accidents or dangerous occurrences
- to require premises or the scene of an accident to be left undisturbed
- to take samples of suspect substances, etc. (they must leave an equivalent sample for independent analysis should you wish it)
- to dismantle or test any dangerous article or substance
- to take possession of any dangerous article or substance for examination or for use in legal proceedings
- to require information facilities and assistance to carry on these duties
- to require the production of any relevant books or documents
- to seize, destroy or render harmless any article or substance which is a source of imminent danger
- to interview employees, take statements and require a written declaration of the truth of these statements.

The following offences carry a maximum fine on summary conviction:

- preventing or hindering a person appearing before an inspector
- intentionally obstructing an inspector.

Enforcement notices

The HSWA brought about an innovation in the enforcement of safety legislation by giving inspectors the power to issue improvement or prohibition notices. Each serves a specific purpose.

Improvement notice

Inspectors may serve an improvement notice where they are of the opinion that a contravention of a statutory requirement has been or is about to be carried out.

This requires the person on whom the notice is served to make the necessary improvements to reach the desired standard. The inspector may specify appropriate remedial action and give a time limit of not less than 21 days in which to comply. The minimum time period is important because it is the time allowed for the person to appeal to an industrial tribunal against the terms of the notice. Appeal may take two forms:

- appeal on the grounds that the notice and remedial action is not reasonable
- the time allowed is not sufficient in which to comply.

When an appeal is lodged it has the effect of suspending the notice until the appeal is heard and decided.

It is interesting to note that an improvement notice cannot be served where the breach has already taken place and is not likely to be repeated. Under these circumstances the inspector may choose to prosecute.

Prohibition notice

A prohibition notice can be served where an inspector is of the opinion that an activity being carried on or about to be carried on may give rise to serious personal injury. There does not have to be a breach of a statutory requirement.

The notice takes effect immediately but unlike the improvement notice an appeal (within 21 days) does not have the effect of suspending the notice.

A failure to comply with either notice is an indictable offence and will result in prosecution with severe penalties. For example, on conviction, a person shall be liable to an unlimited fine and/or imprisonment for a term not exceeding two years.

The lesson here is not to ignore a notice once it has been served just because you disagree with it. This is an offence in itself even if the notice is unreasonable! The best approach under these circumstances is to lodge an appeal and argue the issue of reasonableness in front of a tribunal.

Prosecution

An inspector may choose a third course of action and that is to prosecute the person responsible for committing the offence. This person may be the employer, the employee, individual managers and directors, or any other person assigned responsibilities by the HSWA.

Summary offences

- contravention of a requirement imposed by an inspector under section 20 or 25
- preventing a person from answering an inspector's question
- obstructing an inspector from carrying out his/her duties

- contravening a requirement imposed by a committee of inquiry or to obstruct a person acting under its direction.

The main factors to be considered when deciding whether reasonable care has been taken are the magnitude of the risk involved in the activity and the cost of the precautions necessary for averting the risk.

Other factors to be considered are the obvious nature of the risk and also whether or not the risk is inherent in the job, for example steeplejacks, miners, etc.

PERSONAL LIABILITY

Our general duty as employees is covered by section 7 of the HSWA which says that each employee shall take reasonable care for his own health and safety, and that of other people. We are also required to co-operate with our employer to enable him to carry out his statutory duties. This applies to all employees regardless of status, be they operators, managers or directors.

However, if your job involves instructing other people in what they must do then the HSWA places an extra duty on you to avoid causing these people to commit an offence. Section 36 says:

'Where a person by his act or default causes another person to commit an offence then he, as well as that other person may be charged with the offence.'

Hence if you instruct one of your staff to do something which is a contravention of one of the statutory requirements you may be found guilty of the offence. Several managers have fallen foul of this section. The issues relating to senior managers and directors are a little more complex. It is recognised that employers are merely organisations of people and that offences committed by the organisation may have been due to the act of one or more of its senior managers.

Section 37

'Where an offence committed by a body corporate is shown to have been committed due to the consent, connivance or neglect of a director or senior manager then he as well as the body corporate shall be guilty of the offence.'

Someone then, must make or neglect to make the decisions that affect the safety and well-being of the organisation and its employees, and that individual must be accountable in law.

In a case arising from a serious accident in which the injured party lost four fingers when they became trapped in the rolls of a copying machine, section 37 was used to prosecute the senior office manager. He was given the responsibility of checking the safety of the machine guards by his company's safety policy.

The HSWA has therefore provided a vehicle for the prosecution and punishment of individuals who break the safety laws and put themselves and others at risk.

An important consideration, along with the responsibilities placed by the HSWA, is the extent and domain of responsibility given to individual managers and directors by the organisation's safety policy. The purpose of assigning these responsibilities is to maintain standards of safety in the workplace by giving individuals specific tasks relating to the safe performance of their activities.

In fact where these responsibilities are clearly defined by the policy the individual may incur criminal liability for failing to discharge them.

A director was prosecuted following a fatal accident to one of his employees on the evidence that he had failed to perform a duty of which he knew or ought to have known. He had, in fact, failed to comply with a company circular requiring him to devise a safety policy for his department and to inform and train his employees in the requirements of the HSWA.

More recently the first ever prison sentence was imposed upon a director of a company who ignored a prohibition notice issued by an HSE inspector. The court said 'You have demonstrated a cynical disregard for the dangers to health. I am satisfied your motive was profit'. The prison sentence of one month was suspended for two years.

Safety then, is the employer's responsibility, but is clearly seen as the responsibility of the management in any organisation or business.

Managerial responsibilities

Typical managerial duties set down in a safety policy may include any or all of the following:

- to ensure that you and your staff understand and implement the organisation's health and safety policy
- to ensure that you and your staff are aware of their legal and common law duties relating to health and safety at work
- to ensure that any special regulations which apply to operations carried out under your control are made known and observed, and to see that all procedures are properly carried out
- to see the job safety requirements are listed and made known to operatives by satisfactory safety training
- to assist your staff in resolving health and safety problems and seek assistance as appropriate from the organisation's safety officer
- to ensure that you and members of your staff attend regular meetings at which health and safety can be discussed
- to make sure that the reporting of all accidents and dangerous occurrences is done in order to enable investigations to be carried out and to accept the advice given to prevent recurrence

- to ensure that there is provided adequate levels of supervision suitably qualified and knowledgeable on safe working procedures and practices
- to ensure that all employees within your control are provided with such information, instruction and training to enable them to work safely within your organisation as well as statutory safety requirements, both at the induction stage of employment and during all employment activities
- to ensure that all required safety equipment is made available for use by those employees who have a requirement to use
- to conduct periodic safety inspections or contribute to safety inspections of the area for which you are responsible and to implement improvements as required. This management involvement is crucial to the success of the safety policy.

When judging your safety record you probably looked at the number of accidents you have had over a specified period of time, and compared this with the organisation's accident rate. If you used this measure you might relax in the knowledge that your department has relatively few accidents. However, while this may be comforting news a more important question to be answered is 'how much is this good accident record related to the amount of activity you have undertaken in promoting and maintaining safety standards?' See Chapter 8.

Employee motivation

- How many employees were contacted personally about safety matters?
- How many positive reinforcements of behaviour were given?
- How many negative reinforcements of behaviour were given (reprimands or corrections)?
- What media have been utilised for this purpose, for example, films, posters, letters, etc?

Your personal contact with employees is important in generating an awareness of health and safety. The more interest you show the more likely employees are to co-operate with you in promoting health and safety.

Safety is the responsibility of management and should be managed the same as any other activity within the organisation. However, whilst management normally accepts the responsibility of dealing with matters such as production, industrial relations and the utilisation of plant and equipment etc., they are generally not so careful about ensuring compliance with the safety rules. This is often because the internal penalties for failure in this area are not so clearly defined – managers rarely get sacked for a poor safety record. This should not, however, encourage you to relax your activities in this area.

Selection of contractors

All the effort you put into providing for the safety of your staff may be undone by the activities of contractors in your workplace. It is important to:

- take reasonable care to select competent contractors
- supervise the activities of contractors where their work brings them into contact with your employees.

The common law places an obligation on occupiers of premises to take reasonable care in the selection of contractors. If part of your job includes the hiring of contractors to carry out certain activities this responsibility falls on you. You may be helped in this by your organisation's policy on the employment of contractors. Many specific organisations now carry lists of approved contractors who can be relied upon to work to an acceptable standard. These contractors must be used whenever possible.

In the absence of clear guidelines you should ensure that, in the right circumstances, they have a safety policy outlining the arrangements and procedures their companies have made for safe working.

Where the contractor is required by law to have a safety policy you should ask for it to be presented with the tender for the contract. If the contractor has less than five employees they are not required to have a written safety policy and therefore should be required to give assurances about safety when they are carrying out the work.

These contractors should be required to work in accordance with your organisation's safety policy and given information, instruction and training where necessary to enable them to carry out their duties safely. Where there is uncertainty you may find it helpful to consult your organisation's safety adviser.

Supervision and monitoring

Where the work is carried out in an area under your control you are responsible for ensuring that your staff and members of the public are not put at risk. You will be faced from time to time with decisions relating to the work of contractors.

Permission to start work

All contractors should report to the manager of the affected area before work is commenced. (There will also be a requirement for them to report to the clerk of works or engineering department for longer contracts.)

Use of organisation equipment

Contractors are expected to provide their own tools, plant and equipment necessary for the satisfactory performance of the work in hand. Use of the organisation's equipment should be by written permission of the organisation's representative.

Contractors should also ensure that their employees are provided with any necessary personal safety equipment, for example helmets, goggles, etc.

Fire extinguishers

Contractors should familiarise themselves with the location of fire extinguishers in the area in which they are working. If additional extinguishers are required these should not be obtained from other locations, but from the engineering department.

Fire alarm systems
Contractors should make themselves aware of the position of the fire alarm call points and the organisation's fire procedure.

Interference with fire equipment
Except for use in an emergency, any interference with the electrical wiring and other mechanism of automatic fire detection and extinguishing apparatus, or of the audible fire warning system should be prohibited. Requests for temporary resetting of such equipment must be made to the engineering department.

No smoking areas
The no smoking rules in all areas where smoking is forbidden should be strictly observed. Consideration must also be given to the use of welding equipment, blow lamps, etc., and also to the safe storage and use of flammable materials.

Warning of danger
Care must be taken at all times to protect the organisation's employees, property and work in progress from danger, and any circumstances which give rise to such danger must be reported immediately.

Accidents
All accidents, injuries and dangerous occurrences should be reported to the manager.

Unused materials
Contractors should be required to remove all unused materials and leave the site clean and tidy on completion of the work.

Housekeeping
Contracts should ensure that the workplace is kept tidy with no dirt or refuse being allowed to accumulate.

These points are relevant guidelines to be applied to contractors where their work interfaces with the activities of your employees. Other, more technical, aspects that need to be considered which take into account the needs of both employees and members of the public include:

- *electricity, gas, air mains* – on no account should use be made of the organisation's electricity, gas or compressed air mains, without the permission of the organisation. Where permission is granted the method of connection should be approved by the engineering department

- *excavations* – before any excavation work is begun, the architects or technical services should be consulted about the existence of electric cables, drains, air, gas, and water mains
- *fencing excavations* – excavation or projecting equipment should not be left at any time without taking the necessary safety precautions. All excavations, and openings, should be securely fenced and these and any obstructions marked by a sufficient number of lamps during hours of darkness. During the excavation of all such work, the surrounding area should be maintained in a state of tidiness. Loose materials of whatever kind should not be left about or allowed to obstruct roadways, gangways or working areas
- *building operations* – all building work should be carried our strictly in accordance with the regulations made under the Factories Act in particular complying with the requirements for scaffolding and access equipment.

These provisions, though not exhaustive, may form the basis of a supervision checklist which you might use to assess the activities of contractors.

Remember, just because they are not your employees does not mean to say that you have no responsibility for them. Many of these duties are interrelated but in principle it is the responsibility of the party in control of, or carrying out, the activity to foresee any possible harm and take reasonable steps to prevent it. This is complicated by the fact that some areas of responsibility may be shared by two or more parties so that clear procedures need to be laid down to avoid confusion. This is particularly so where contractors are working alongside your own employees and members of the public.

Above all there is a need to ensure that all parties are aware of the risks involved by providing information instruction and training, and by ensuring that their activities are supervised and monitored.

By considering all these points you will not only help to ensure the safety of employees and others, but also help to avoid incurring liability on both yourself and your employer.

ACTS OF PARLIAMENT

Legislation has consolidated the common law rules of employers liability and has given rise to a civil liability to pay compensation.

The Employers Liability (Defective Equipment) Act 1969

This Act made the employer liable for the fault of the manufacture of defective equipment. The employer will have to pay the employee compensation. He will then be able to claim his money back from the manufacturer. This is known as the principle of loss distribution.

Employers' Liability (Compulsory Insurance) Act 1969

The Act makes it obligatory for most employers (except nationalised industries, local authorities and the police) to insure against liability for injury to employees. Failure to buy such insurance or display a certificate of insurance can lead to fines on a daily basis of up to £500 and £200 respectively under the Criminal Law Act 1977.

Occupiers' Liability Act 1959

This imposes duties on occupiers of premises. Where an employee goes to work on third party premises, his employer is liable if the employee is injured through the employer having failed to set up a safe system of work. However, if the employee is injured by a structural defect in the premises which is not incidental to the work, the employer is not liable. The injured person must sue the occupier of the premises. This will only succeed if there are unusual dangers which the occupier knew or should have known about. Until the Unfair Contract Terms Act 1977, it was possible for the occupier to contract out of their liability by posting a warning notice and condition.

Congenital Disabilities (Civil Liability) Act 1976

This Act makes it possible for a child, who is born disabled as a result of a breach of common law or statutory duty to a parent, to sue the parent's employer.

Fatal Accidents Act 1976

This allows dependents of employees killed at work through the employer's negligence, to sue the employer for damages. Compensation is assessed on the actual or prospective earnings of the deceased breadwinner.

Law Reform (Miscellaneous Provisions) Act 1934

This allows the person entitled to the property of a deceased employee killed at work, to inherit any right of action against the employer which the deceased person might have had.

Limitation Act 1975

This controls the time period within which actions must be brought against the employer. The period is normally three years from the date when the cause of the action began. The cause of action begins when the injured employee had the necessary legal and medical knowledge to realise that he had a probable right to sue his employer.

Some other relevant legislation is given in Table 2.1 in the form of a checklist. Whilst this might not be exhaustive you should enlarge upon this list and attempt to keep it up to date. A blank checklist is provided for this purpose in Table 2.2.

Table 2.1 Some Acts of Parliament

Legislation	Does it apply to you?		Have you a copy?		Do you understand it?	
	YES	NO	YES	NO	YES	NO
Health and Morals Act 1833						
Factory Act 1833						
Mines and Collieries Act 1842						
Workshop Regulations Act 1967						
Metalliferous Mines Regulation 1872						
Explosives Act 1875						
Employers' Liability Act 1880						
Shop Hours Regulations Act 1886						
Factory and Workshop Act 1901						
Alkali Works Regulations act 1906						
Coal Mines Act 1911						
Anthrax Prevention Act 1919						
Employment of Women, Young Persons and Children Act 1920						
Celluloid and Cinematograph Film Act 1922						
Explosives Act 1923						
Petroleum Consolidation Act 1928						
Children and Young Persons Act 1933						
Hours of Employment (Conventions) Act 1936						
Petroleum (Transfer of Licences) Act 1936						
Public Health Act 1936						
Children and Young Persons (Scotland) Act 1937						
Young Persons (Employment) Act 1938						
Law Reform (Contributory Negligence) Act 1954						
Ministry of Fuel and Power Act 1945						
Coal Industry Nationalisation Act 1946						
Radioactive Substances Act 1948						
Factories Act 1948						
Law Reform (Personal Injuries Act) 1948						
Shops Act 1950						
Rag Flock and Other Filling Materials Act 1951						
Rivers (Prevention of Pollution) Act 1951						
Fireworks Act 1951						
Agriculture (Safety, Health and Welfare Provisions) Act 1956						
Clean Air Act 1956						
Occupiers' Liability Act 1957						
Factories Act 1959						
Radioactive Substances Act 1960						
Factories Act 1961						
Offices, Shops and Railway Premises Act 1963						
Nuclear Installations Act 1965						
Civil Evidence Act 1968						
Trade Descriptions Act 1968						
Employer's Liability (Compulsory Insurance) Act 1969						

Table 2.1 (Continued)

Legislation	Does it apply to you?		Have you a copy?		Do you understand it?	
	YES	NO	YES	NO	YES	NO
Employer's Liability (Defective Equipment) Act 1969						
Public Health (Recurring Nuisances) Act 1969						
Equal Pay Act 1970						
Fire Precautions Act 1971						
Industrial Relations Act 1971						
Mines Management Act 1971						
Employment Medical Advisory Service Act 1972						
Control of Pollution Act 1974						
Health and Safety at Work Act 1974						
Trade Union and Labour Relations Act 1974						
Water Resources Act 1974						
Limitations Act 1975						
Sex Discrimination Act 1975						
Social Security Act 1975						
Social Security (Consequential Provisions) Act 1975						
Fatal Accidents Act 1976						
Food and Drugs (Control of Premises) Act 1976						
Race Relations Act 1976						
Trade Union and Labour Relations (Amendment) Act 1976						
Criminal Law Act 1977						
Employment Protection (Consolidation) Act 1978						
Refuse Disposal (Amenity) Act 1978						
Employment Act 1980						
Limitation Act 1980						
Magistrates Court Act 1980						
Local Government, Planning and Land Act 1980						
Administration of Justice Act 1982						
Derelict Land Act 1982						
Social Security and Housing Benefits Act 1982						
Supply of Goods and Services Act 1982						
Criminal Justice Act 1982						
Food Act 1984						
Road Traffic Regulation Act 1984						
National Health Service (Amendment) Act 1986						
Safety at Sea Act 1986						
Sex Discrimination Act 1986						
Fire Safety and Safety of Places of Sport Act 1987						
Consumer Protection Act 1987						
Road Traffic Act 1988						
Local Government Act 1988						
Road Traffic (Driver Licensing and Information Systems) Act 1989						
Companies Act 1989						
Environmental Protection Act 1990						

Table 2.2 Acts of Parliament update

Legislation	Does it apply to you?		Have you a copy?		Do you understand it?	
	YES	NO	YES	NO	YES	NO

REGULATIONS

Regulations proposed by the HSC, accepted by the Secretary of State and duly laid down before Parliament become the law of the land and are duly enforceable through the courts. Breaches of duties imposed by health and safety regulations (where they cause damage) are actionable in civil proceedings unless they specify otherwise. For example, the HSWA specifically excludes sections 2–9 from being actionable at common law for breach of a statutory duty. New regulations may place an absolute duty on employers in the same way as some existing provisions do, or they may be limited by the use of phrases such as 'so far as practicable' and 'so far as reasonably practicable'. 'Reasonably practicable' on the other hand enables account to be taken of differing circumstances.

Where the word 'practicable' is used it is generally taken to mean that which is possible in the light of current knowledge and invention. For example, it is impossible to take precautions against a danger which is not known to exist, or to take precautions which have not yet been invented.

Approved codes of practice

The HSC's power to approve the issue codes of practice is contained in section 16 of the HSWA. The objective of these codes is to provide practical guidance on the requirements contained in sections 2–7 of the Act, or in health and safety regulations, or in any of the existing statutory provisions. The precise statutory requirement to which the approved code is supplementary must be indicated in the code itself.

The purpose of a code of practice is to:

- specify in greater detail or in a more liberal style than is practicable or desirable in regulations, the precise technical and other requirements to be observed in a particular set of regulations or other relevant statutory provisions.
- explain what in particular circumstances would be considered to constitute satisfactory compliance with the requirements of a general obligation.

It was the intention of Parliament that an approved code of practice should enjoy a new and special status and would in effect provide a simple and flexible extension of the law. This special status is given by section 17 which states that failure to observe any provision of an approved code shall not of itself render a person liable to criminal or civil proceedings, but where in criminal proceedings a person is alleged to have contravened a statutory requirement or prohibition, the court is required to admit in evidence any provisions of an approved code which appear to it to be relevant. Further, where the court considers that provisions of the code are relevant, then failure to observe those provisions is to be taken as proof of a contravention unless the court can be satisfied that you have complied with the requirement in another way.

Approved codes will usually be generated by the HSC, the HSE, industry advisory committees set by up the HSC, or possibly in conjunction with such a body as the British Standards Institute. Codes generated elsewhere will not normally be approved by the HSC unless they are felt to be a necessary and suitable extension of the law.

Reasonably practicable

An illustration of the term 'reasonably practicable' is to be seen in the case of *Associated Dairies v. Hartley ([1979] 1 RLR 175)*. A worker was injured when a wheel on a lorry ran over his toe. The employer had provided safety shoes which could be purchased at cost by the workforce. The cost was spread over a period of time. An inspector issued an improvement notice requiring Associated Dairies to issue the shoes free of cost. The employer appealed. The tribunal thought that it would be practicable for the employer to provide the shoes without cost, but that it was not reasonable to expect him to do so. The time, trouble and expense of safety precautions must be weighed against the foreseeable risk and in this case, the expenditure by the employers would be disproportionate to the risk.

Offences under the Act and those other safety and health laws mentioned are deemed criminal and may involve prosecution of an individual or a company in a Magistrates' or Crown court. It is a fundamental principle of criminal law for the prosecutor to prove beyond reasonable doubt that the offence was committed by the accused. Under section 40 of the HSWA, it is for the accused to prove that he did take reasonably practicable precautions as required by the Act and by some sections of the other health and safety laws. This means that if the accused thought that it was not reasonably practicable to take certain steps then it is for him to provide the proof.

A checklist of some of the relevant regulations in force is given in Table 2.3. You should attempt to keep this up to date and a blank checklist is given in Table 2.4.

Table 2.3 Some statutory instruments, rules and orders

Some statutory instruments, rules and orders	Does it apply to you?		Have you a copy?		Do you understand it?	
	YES	NO	YES	NO	YES	NO
Factories, Locomotives and Sidings Regulations 1906						
Factories, (Horsehair Processes) Regulations 1907						
Factores (Electrical Energy) Regulations 1908						
Factories (Tin or Terne Plates Manufacturing) Order 1917						
Tanning (Two-Bath Process) Welfare Order 1918						
Fruit Preserving (Welfare) Order 1920						
Laundry Workers (Welfare) Order 1920						
Celluloid (Manufacturing) Regulations 1921						
Chemical Works Regulations 1925						
Herring Curing (Scotland) Welfare Order 1926						
Bakehouses Welfare Order 1927						
Manufacture of Cinematograph Film Regulations 1928						
Oil Cake Welfare Order 1929						
Cement Works Welfare Order 1930						
Sugar Factories Welfare Order 1931						
Sanitary Accommodation Regulations 1939						
Cinematograph Film Stripping Regulations 1939						
Electricity (Factores Act) Special Regulations 1950						
Factories (Testing of Aircraft Engines) Regulations 1952						
Iron & Steel Foundries Regulations 1953						
Petroleum Spirit (Conveyance by Road) Regulations 1957						
Agriculture (Avoidance of Accidents to Children) Regulations 1958						
Agriculture (Safeguarding of Workplaces) Regulations 1959						
Agriculture (Threshers and Balers) Regulations 1960						
Construction (General Provisions) Regulations 1961						
Agriculture (Field Machinery) Regulations 1964						
Washing Facilities Regulations 1964						
Examination of Steam Boilers Regulations 1962						
Power Presses Regulations 1964						
Construction (Working Places) Regulations 1966						
Construction (Health and Welfare) Regulations 1966						
Carcinogenic Substances Regulations 1967						
Offices, Shops and Railway Premises (Hoists and Lifts) Regulations 1968						
Asbestos Regulations 1969						
Abrasive Wheels Regulations 1970						
Foundaries (Protective Footware and Gaiters) Regulations 1971						
Highly Flammable Liquids and Petroleum Gases Regulations 1972						
Organic Peroxides (Conveyance by Road) Regulations 1973						
Agriculture (Tractor Cabs) Regulations 1974						
Industrial Tribunals (Improvement and Prohibitions) Regulations 1974						
Woodworking Machines Regulations 1974						
Protection of Eyes Regulations 1974						

Table 2.3 (Continued)

Some statutory instruments, rules and orders	Does it apply to you?		Have you a copy?		Do you understand it?	
	YES	NO	YES	NO	YES	NO
Employers' Health and Safety Policy Statements (Ex) Regulations 1975						
Protection of Eyes (Amendment) Regulations 1976						
Safety Representatives and Safety Committees Regulations 1977						
Health and Safety (Enforcing Authority) Regulations 1977						
Motor Vehicles (Constructions and Use) Regulations 1978						
Control of Lead at Work Regulations 1980						
Safety Signs Regulations 1980						
Dangerous Substances (Conveyance by Road) Regulations 1981						
Diving Operations at Work Regulations 1981						
Health and Safety (Dangerous Pathogens) Regulations 1981						
Health and Safety (First-Aid) Regulations 1981						
Notification of Installations Handling Hazardous Regulations 1984						
Criminal Penalties (Increase) Order 1984						
Classification, Packaging and Labelling of Dangerous Substances Regulations 1984						
Poisonous Substances in Agriculture Regulations 1984						
Social Security (Industrial Injuries) (Prescribed Diseases) Regulations 1985						
Ionising Radiation Regulations 1985						
Industrial Tribunals (Rules of Procedure) Regulations 1985						
Building Regulations 1985						
Electrically Operated Lifts (EEC Requirements) Regulations 1986						
Dangerous Substances in Harbour Areas Regulations 1987						
Control of Asbestos at Work Regulations 1987						
Control of Substances Hazardous to Health Regulations 1988						
Pneumoconiosis (Workers Compensation) (Specified Diseases) Regulations 1988						
Ionising Radiations (Protection of Persons Undergoing Medical Examination or Treatment) Regulations 1988						
Classification, Packaging and Labelling of Dangerous Substances (Amendment) Regulation 1989						
Road Vehicles Lighting Regulations 1989						
Freight Containers (Safety Convention) Regulations 1989						
Road Traffic Accidents (Payment for Treatment) Regulations 1989						
Social Security (Industrial Injuries and Disease) Misc Provisions 1990						
Collision Regulation (Seaplanes) (Amendment Order) 1990						
Dangerous Substances (Notification and Marking of Sites) Regulations 1990						
Smoke Control Areas (Authorised Fuels) Regulations 1990						
Personal Injuries (Civilians) Regulations, 1990						
Cosmetic Products (Safety) (Amendment) Regulations 1990						
Fire Safety and Safety of Places of Sport Act 1987 (Commencement Order No 6) Order 1990						

THE OFFICE SAFETY HANDBOOK

Table 2.4 Statutory instruments, rules and orders update

Some statutory instruments, rules and orders	Does it apply to you?		Have you a copy?		Do you understand it?	
	YES	NO	YES	NO	YES	NO

THE EUROPEAN UNION

Since the formation of the European Union, it must not be forgotten that the European Parliament may issue directives which will influence policies and strategies within member states. Article 118A of the Treaty of Rome provides that the European Council shall adopt, by means of directives, minimum requirements for encouraging improvements regarding the health and safety of workers, especially in the working environment. Article 15 of this directive states that member states will have brought into force the laws, regulations and administrative procedures necessary to comply with this directive by 1 January 1991. The directive covers the following areas:

- objectives
- definitions
- responsibilities of employers
- obligations of the employer
- preventative services
- information to be collected and held by employers
- information and communication with workers
- consultation with the workforce
- training
- obligation placed on workers
- adoption of individual directives concerning the workplace, equipment, personal protective equipment, work and visual display units and the handling of heavy loads involving the risk of back injury.

Whilst the law places statutory obligations upon organisations such as those described above it is important that these are implemented efficiently and effectively. Subsequent chapters will detail how this can be achieved.

SUMMARY

It is not possible here to provide an in-depth study into the health and safety legislation in this country. It is only intended to remind those already familiar with the law and to provide an introduction to the lay person or student. A reading list is provided for those wishing to learn more about this complex area. You only need a working knowledge of the law as would any professional manager. Expert advice should be sought as and when appropriate owing to the complexities of the law in this regard. Despite the wealth of safety case law in existence at the present time, it is always a communication problem in getting the 'lessons learnt' aspects of the cases across to safety practitioners. Many safety practitioners experience great difficulty in this area. It is essential, therefore, that you have direct access to legal expertise. At the present time, there is a bewildering mass of legislation which lays both criminal and civil responsibilities upon employers. Amongst aspects covered are the style and manner of accident reporting, the safe access to buildings and sites and particular regulations governing dangerous materials or processes such as asbestos and steam boilers. Special constraints are placed on the employment of children and young persons. Construction and building operations because of their high risk nature are highly regulated. Procedures for the control of industrial major accident hazards and the use and transportation of dangerous substances have all been introduced in the 1980s.

Exceptional care has to be taken in handling the relationship between safety and discipline and unfair dismissal even though employees' duties are tightly defined. Distinct regulations apply to electrical and fire hazards within the Factories Act. Specific provision has to be made for appropriate first-aid facilities. It must be stated that with regard to enforcement, draconian consequences both for civil and criminal liability may be felt by board members down to low level supervisors. Specific controls relating to food handling and hygiene, use of lifting machinery and materials' handling are all covered. Chronic problems which may be felt by employees in terms of noise and vibration, overcrowding or problems of occupational health have all been strengthened in the last three years. Liability may also exist for product and structural safety as well as personal protection. Certain categories of workers including women and those who work at heights or offshore are afforded additional protection under the law.

In the UK as in Europe, regulations, directives or decrees are introduced from time to time to deal with issues which have been found to be dangerous or hazardous to health. These regulations might cover such subjects as the use of circular saws or washing facilities. They are usually introduced following evidence to the enforcement agencies that such guidance is necessary. This is an established practice in the western world and covers all aspects of safety. For example, evidence was provided that alcohol was a contributory factor in road accidents. Most developed countries have legislated, or issued decrees or regulations, which clearly point out the consequences of drinking alcohol and driving following evidence gained from accident analysis. In the same way industrial enforcement agencies gather similar evidence which assists in introducing appropriate measures designed to reduce or prevent such occurrences happening in the future. Senior management must appreciate that the rules are made to help not hinder the organisational mission. On the other hand regulating against some malpractice does not necessarily prevent it from happening. People still injure themselves on chain saws, still drink and drive and still suffer ill health carrying on some activity contrary to good advice or legal necessity.

Finally, the law must be regularly monitored. It is important to establish as early as possible the implication of each set piece of legislation so that appropriate management decisions can be taken. Keeping abreast of current legislation and/or regulations can be time consuming and it is important to know where to seek advice quickly and easily. This will require the establishment of systems and practices designed specifically to inform organisations of their legal obligations at regular intervals. This could include attendance at training courses, seminars and conferences which may be of a national or in-house style. The implication of new regulations upon an organisation are important. The question most likely to be asked a safety manager is in relation to the cost to the organisation. This will require a detailed reply. The procedure for doing this must form part of the daily management plan. These issues are discussed in subsequent chapters dealing with safety management planning and implementation.

With industrial accidents increasing each year it is clear that the law alone is unable to control this epidemic effectively. For effective accident reduction strategies we require a co-ordinated strategy which some refer to as the safety mix or 4 E strategy. This refers to *environmental* strategies, such as making the workplace safer or healthier; *engineering* by making machinery and tools safer, *education* programmes designed to raise the awareness of all problems in the workplace; and *enforcement* which covers the law in this regard. The management of safety must be seen as a corporate issue which demands the skills of properly trained safety managers (or office managers/supervisors where an organisation is unable to employ a qualified safety practitioner) who are capable of utilising all the resources with an organisation. They must have a sound understanding of latest developments within the 4 E philosophy and have sufficient authority within the organisation to implement sound proactive policies.

ACTION CHECKLIST

- Do you understand the difference between civil and criminal liability?
- Do you understand the duties and responsibilities of employer and employee under the terms of the HSWA 1974?
- Do you understand your responsibilities under the current health and safety legislation. For example, does COSHH apply to you?
- Do you have a system for keeping up to date with current legislative requirements in this country and those under EU directives?
- Do you have access to appropriate technical and professional expertise?
- Do you know where to get help and advice?
- Do you have appropriate management systems in place to ensure that your organisation conforms to the law?
- Do you know whether your organisation is meeting its full obligations in law at this moment?

'... prepare and keep up to date, a written statement of your policy regarding health and safety ...'

How to implement your safety policy

SAFETY POLICY STATEMENT

One of the primary legal provisions of section 2 of the HSWA requires employers who employ five people or more to prepare, and keep up to date, a written statement of their policy regarding the health and safety of their employees.

Your policy may consist of one complete manual or it may be a compilation of several individual documents relating to particular areas of your activity. Whatever form it takes, it should comprise three separate and distinct parts:

- general statement of policy
- organisation and responsibilities for carrying out the policy
- arrangements for ensuring safety and health of employees.

Many organisations produce a general statement of intent and require individual departments, cost centres or directorates and/or offices to develop safety policies which appropriately reflect the day-to-day activities of the department as a whole.

By looking through the policy documentation, you should find an indication that the organisation will provide resources for health and safety along with the provision of safe plant and equipment, safe systems of work, training for staff and supervision. In fact, you should observe a resemblance to the general duties laid down by section 2 of the HSWA.

As far as responsibilities are concerned, these should include the duties of employees at all levels in the organisation.

Section 1 General Statement of Policy

Here you should state policy in respect of:

- provision and maintenance of safe and healthy working conditions

- equipment and safe systems of work for all employees
- provision of health and safety information, training and supervision necessary
- a statement accepting responsibility for the health and safety of other people who may be affected by your activities.

Within this section, it is good practice to outline the allocation of duties for all safety matters and the particular arrangements which will be made to implement your company health and safety policy. Your statement of health and safety policy should always reflect the changes in nature and size of the company and as such be kept up to date. To ensure this, you should see to it that such statements are reviewed every year and amended accordingly.

Section 2   Organisation and responsibilities for carrying out the policy

This section should cover those general arrangements for carrying out health and safety policy such as:

- procedures for the reporting and recording of accidents and dangerous occurrences
- location of first-aid boxes
- details of all qualified first-aid personnel
- details of all fire procedures
- fire extinguisher location and maintenance details
- fire alarm location and maintenance details
- fire routines and testing arrangements
- medical and health care arrangements
- safety training details including specialist training
- advice for visitors and/or contractors
- discipline and codes of conduct

Section 3   Arrangements for ensuring safety and health of employees

The last section should list details concerning all hazards in the workplace. These should be listed so that everyone is aware of them. If you use hazardous substances, then the manufacturer will provide hazard sheets, whilst the Health and Safety Executive (HSE) will provide additional advice if it is required. Make sure that the workforce is aware of all hazards. This section should also provide information concerning:

- cleanliness of the premises
- waste disposal details

- details concerning safe stacking and storage
- marking and maintaining clear walkways and exits
- equipment checking procedures
- details concerning access to restricted areas
- routines for checking electrical appliances
- routines for the reporting of faults
- rules regarding the use of extension cables and portable equipment
- arrangements with electrical and or equipment contractors
- rules concerning the use of all machinery
- details of routine maintenance and timetable
- dangerous substance details
- protective equipment and clothing information
- routines for storing, handling and disposal of dangerous substances
- operation, use and maintenance of compressed air equipment
- storage, labelling and use of any compressed gases and or fluids procedures regarding water pressure and or steam
- in-house rules and regulations such as:
  - use of internal transport
  - use and care of protective equipment and clothing
  - noise
  - maintenance of appliances
  - any other special hazards in your company office.

It is good practice to include a company's health and safety policy in the induction course programme for new employees. A sample health and safety policy document is given in Figure 3.1.

SAFETY REPRESENTATIVES AND COMMITTEES

A safety representative is an employee nominated by his trades union to represent his colleagues in discussions with the employer on matters relating to health and safety at work. He/she may carry out surveys or inspection of the workplace with the object of identifying hazards or potential dangers. Employers are required to disclose to such representatives all information necessary for them to carry out their duties and functions. Details of the information can be found in para. 6 of the Approved Code of Practice on Safety Representatives and Safety Committees. Health and safety inspectors are also required to provide safety representatives with technical information obtained as a result of a visit to their workplace. This must include details of prosecutions, improvement notices, prohibition orders, and any correspondence to the employer on health and safety matters. All action

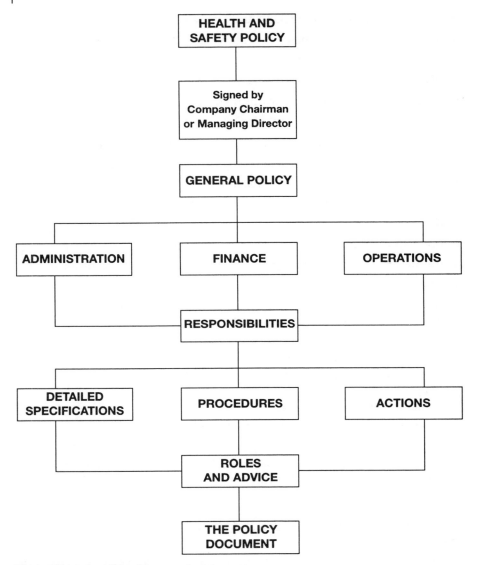

Figure 3.1 A policy document sequence

considered by an inspector as the result of a visit must be discussed with the safety representative.

As far as the office manager is concerned, the safety representative/s within an organisation are an extremely valuable asset. Meet them regularly, involve them directly within the decision-making process and consider them as an extension to the overall health and safety programme within your organisation.

For safety committees to be effective they must not be too large otherwise their ability to make decisions effectively is diminished. They must be made up of senior management, trades union safety representatives with the secretariat being provided for by the safety manager or one of his members of staff. Chair of the committee should follow normal electoral procedure and a constitution should allow for periodical changes to the structure to take place. Decisions taken by the committee must be adhered to and carried out as soon as is reasonably practicable. Merely having a safety committee because the regulations say you have been requested to have one is a severe waste of resources if it is not permitted to make an effective contribution to the well-being of the organisation.

Decisions concerning new laws and regulations affecting working practices should be placed before the safety committee for discussion. From this, all matters concerning the planning, implementation, monitoring and evaluation of policy can be discussed openly and widely. These and other issues are discussed in Chapter 5.

These regulations came into force on 1 October 1978 and gave trades unions the right to appoint safety representatives to perform the following functions:

- to investigate potential hazards and dangerous occurrences at the workplace and to examine the cause of accidents
- to investigate complaints by employees in relation to health, safety and welfare at work and to make representations to the employer on these issues
- to take up general matters of health and safety at work and to represent employees in consultation with HSE inspectors
- to attend meetings of the safety committee when and if necessary
- to carry out a workplace inspection every three months after giving reasonable notice to the employer of intention to do so
- to carry out workplace inspections (after consultation with the employer) when there has been a change in conditions of work, or new information has been issued concerning relevant hazards. The employer is entitled to be represented at these inspections if they wish.

The idea of workers being involved in these activities has been debated for many years and was actually resisted by many trades unions because they suspected that such involvement might relieve employers of some of the responsibility to health and safety.

It is important to note therefore that the activities outlined above are merely functions which the representative may carry out. They do not relieve the employer of any responsibility whatsoever and do not confer any legal liability on the appointed representative.

The regulations specify a number of other provisions relating to the training of safety representatives and the provision of information relating to safety and health at work. Your organisation's safety policy may specify the local arrangements made to enable safety representatives to perform these functions.

Safety representatives are an important influence in the provision and maintenance of a safe and health working environment.

ACCIDENTS AND DANGEROUS OCCURRENCES

Although these are currently under review, the present requirement is to report injuries, diseases and dangerous occurrences as contained in the *Reporting of Injuries, Diseases and Dangerous Occurrences Regulations 1985*. These are sometimes referred to as RIDDOR for short. These regulations require that when any of the following events occur, the responsible person must report it in writing to the enforcing authority and must keep a record of it. The regulations also stress that the responsible person should also notify the enforcing authority by telephone as soon as is reasonably practicable where:

- a death occurs to any person, whether or not he or she is at work, as a result of an accident arising out of or in connection with work
- any person suffering a specified major injury or condition as a result of an accident arising out of or in connection with work
- one of a list of specified dangerous occurrences arising out of or in connection with work
- a person at work being incapacitated for his or her normal work for more than three days as a result of an injury caused by an accident at work
- the death of an employee if this occurs some time after a reportable injury which led to that employee's death, but not more than one year afterwards
- a person at work being affected by one of a number of specified diseases, provided that a medical practitioner diagnoses the disease and that the person's job involves a specified work activity.

A diagram illustrating the present procedure to be followed is given in Figure 3.2.

When making a report, it is necessary to complete Form 2508 (revised January 1986) for the reporting of deaths, injuries and dangerous occurrences and Form 2508A should be used when reporting cases of disease.

The regulations also require employers to keep records of every event which has caused injury, however small, or which has the potential to cause injury. The details to be kept are as follows:

- date and time of the accident or dangerous occurrence
- details of the person(s) affected and include:
  - full name
  - occupation
  - nature of the injury or condition
- location where the accident or occurrence took place
- a brief description of the incident.

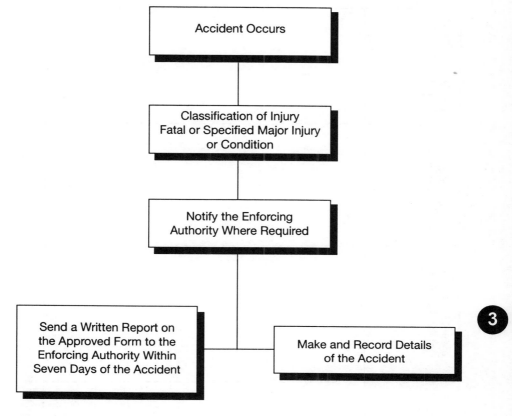

Figure 3.2   How to report and record accidents which happen

In addition, where a disease is concerned then the following information will be recorded:

- date of diagnosis
- personal details of affected person
- name and nature of the disease.

All records must be kept for three years and be made available to the enforcing authority if or when required. There is flexibility in the way records will be kept provided that they contain the above particulars. An example form is given in Figure 3.3.

Date	Time	Location	Number of Injured Persons	Brief Description of the Accident	Action Taken	By Whom Action Taken

Figure 3.3 Internal accident and dangerous occurrence summary sheet

LIABILITY INSURANCE

It is a statutory requirement for most companies in the UK to have an insurance policy which provides cover against claims for injury and/or disease by employees. When a policy is taken out, the insurers provide a certificate of insurance which the employer must display in a prominent position at the work site. This is so that employees and anyone else requiring to see it can do so. It is a criminal offence not to take out such insurance and/or to display the certificate.

The object of having such insurance is to make certain that employers are covered for any legal liability to pay damages to employees who are injured or suffer ill health through their employment. Such a policy protects an employer from third party claims but not for non-employees. Where contractors or members of the public are concerned it is necessary to take out additional insurance referred to as public liability insurance. Whilst such a policy might be desirable it is not a requirement of the law. All statutory liability insurance must be available under one or more approved policies. These policies are not subject to any conditions or exception prohibited by the law.

An employer is not required to insure against liability to an employee who is:

- a father
- a mother
- a son or daughter
- a wife or husband
- a close relative.

With regards to the degree of cover necessary, it is a requirement to maintain insurance of £2,000,000 but most reputable insurance companies have no upper limit. Exempted employers are:

- local authorities
- civil service
- nationalised bodies
- commission for the new towns
- statutory water authorities
- London Transport.

Failure to effect and maintain valid insurance for any day on which it is required is a criminal offence as is failure to display a valid certificate of insurance.

WOMEN WORKERS

There are many books available covering aspects of the Sex Discrimination Act and it is not intended to cover this ground here. In terms of health and safety however, case law has shown that a woman's health and safety in certain circumstances takes precedence over a duty not to discriminate. For example, in the case of *Page v. Freight Hire (Tank Haulage) Ltd [1981] IRLR 13* a 23-year-old woman was employed as a heavy goods vehicle driver of tankers carrying the chemical dimethylformamide (DMF). On a recommendation from ICI that women of child-bearing age should not be employed in driving lorries loaded with the chemical DMF, Mrs Page was taken off this kind of work. She brought a claim of unlawful sex discrimination. The decision did not allow an employer to stop employing women as soon as a hint of danger arose. What needed to be considered was the employer's duty under section 2 of the Health and Safety at Work Act. In this particular case, evidence from the manufacturer was that the chemical was dangerous to women of Mrs Page's age and the steps taken to remove her from the danger was a reasonable one. There was, therefore, no unlawful discrimination.

The effect of the Sex Discrimination Acts of 1975 and 1986 has been to remove some of the restrictions on women in relation to their work but restrictions and prohibitions on certain types of employment are still in force in the interests of health and safety.

INDUSTRIAL TRIBUNALS

Industrial relations is a multi-disciplinary subject, the content of which is still in controversy. This is not surprising because the employment relationship is complex. Essentially it's an economic relationship which takes place within a:

- social
- psychological
- political
- cultural

context, expressed legally in an individual contract of employment. The background to this relationship is often a complex pattern of shifting forces and changing environments, public pressures, private fears, world markets, weekly household bills, driving ambitions and security. The issues involved with industrial relations arise out of who does what? who gets what? and what is fair? in relation to:

- income
- treatment
- time.

These issues provide potential for the emergence of conflict and thus the problems of resolving conflict and its causes are issues of which you must be aware.

Income

Aspects regarded as unfair:

- between the low paid and the rest
- between *earned* and *unearned* income
- between wage earnings, staff salaries and fringe benefits
- between occupations and between industries
- between grades and internal differentials
- between public and private sector
- between jobs in different regions
- does the wage system reflect the changing aspirations of the staff?
- does it accommodate technical change?
- does it enable changes in marketing and sales policies?

Treatment

Are the needs and expectations of the modern industrial worker being catered for in respect of:

- fringe benefits
- staff status
- facilities?

Management needs organisation but workers respond as individuals or as informal working groups. Whilst management might initiate change, the worker may only be interested in security.

Time

- Is overtime necessary?
- Shift work, the conflict between man and machine?
- Stress?
- Bonus schemes?

The time factor has been identified as having a strong association with accident causation.

There are a number of *rule-making* processes regulating the employment relationship. The purpose of rules, both formal and informal, exist to establish 'rights' and 'obligations' which together outline spheres of 'authority' and define 'status' and thus establish 'norms' of expected and appropriate behaviour. The main rules and rule making processes are:

- legislation
- collective bargaining
- unilateral management decisions
- unilateral trades union regulations
- the individual contract of employment
- custom and practice
- arbitration awards both voluntary and statutory
- social conventions.

Thus, apart from the problem of *fairness* providing potential for conflict, so also do aspects of the rule-making processes. For example:

- who should make the rules and how?
- what is the appropriate process?
- what issues should the rules cover?
- how are the rules to be implemented?
- how are the rules to be enforced?
- how are the rules changed?
- how are the rules to be legitimised by those affected?
- what are the effects of new rules on existing managerial perogatives?

These are basically two types of rule:

- *substantive rules* – these include rules governing compensation, rules regulating duties, expected performance and discipline matters. They also cover rules defining rights and duties of employees
- *procedural rules* – are concerned with defining the procedures for the establishment and administration of substantive rules.

In essence, substantive rules define jobs, whilst procedural rules regulate the defining process. Getting agreement on what is regarded as being fair would be the use of job evaluation in order to assess the:

- skill and training requirements
- social worth of the job
- physical effort required
- level of experience and competence required
- level of responsibility
- level of danger and risk.

Factors also relevant are:

- ability and age
- working life expectancy of the job
- ability of the employer to pay
- productivity
- persistence, honesty and reliability
- scarcity value
- social needs and traditions.

Where conflict occurs there must be a means of resolving the matter quickly and safely. If the matter cannot be resolved internally then industrial tribunals have been set up to consider cases where independent arbitration is necessary. They were originally intended to deal with disputes quickly, cheaply and informally to deal with injustices in the employment sphere. To all intents and purposes, the industrial tribunal is made up of a panel of three persons of which the chairman is usually the only one legally qualified. Whilst there is no need to be represented by a legal expert the advice today is that the employment law is so complicated it is often recommended that legal representation or legal aid be provided. Where an employee is a member of a trades union it is usual for the trades union to represent its client.

The make up of a tribunal is usually one trades union panel member, one employer and the chairman who ensures that the complexities of the law are interpreted and followed. For a decision to be reached, a tribunal needs a majority view. Decisions may be challenged provided that:

- the decision discloses an error on the part of the tribunal
- a party did not receive notice of proceedings
- a decision was reached in the absence of a party entitled to be heard
- new evidence has come to light
- the interests of justice require a review.

The chairperson can refuse an application for a review if he/she believes it to be unsuccessful. Applications for reviews must normally be made within 14 days of the decision of the tribunal being made known.

Industrial tribunals are empowered to hear complaints under the following Acts:

- Equal Pay Act 1970
- Health and Safety at Work Act 1974
- Sex Discrimination Acts 1975 and 1986
- Trade Union and Labour Relations Act 1976
- Race Relations Act 1976
- Employment Protection (Consolidation) Act 1978

- The Employment Acts 1980 and 1982

and subsequent similar legislation. Some examples of health and safety matters of interest to the office manager are as follows:

- failure of an employer to pay safety representatives for time off for carrying out duties and for training
- failure of an employer to make a medical suspension payment
- appeals against improvement and prohibition notices
- time off for training of safety representatives
- dismissal, whether actual or constructive, following a breach of health and safety law or regulation, or term of an employment contract.

Appeals are made against improvement notices and prohibition notices on the following grounds:

- substantive law involved, or
- time limits for compliance.

Matters regarding grievance procedures, negotiation, bargaining and the study of industrial relations will be covered in more detail in other books in this series.

ACTION CHECKLIST

- *What plans exist to continue trading if a fire damaged your company premises?*
- *Have emergency plans been tested and how recently?*
- *How up to date is the company statement on health and safety policy?*
- *How is the contents of the office health and safety policy document disseminated to members of staff?*
- *Do we need a health and safety committee?*
- *Do we have a procedure for dealing with accidents and dangerous occurrences?*
- *Is our liability insurance valid, adequate and displayed in an appropriate location?*
- *Do I understand the requirements for the health and safety of female employees?*
- *Have we an appropriate industrial relations mechanism for dealing with matters relating to health and safety?*

‘Gathering good reliable information must apply to the role of office manager’

‘... few exploit the value of such costly information as a management tool’

Accident records

INFORMATION GATHERING

Studies continually find that many organisations do not keep any accident or dangerous occurrence records other than those provided for within the compass of Form F2508. Sometimes even this information is not kept. Many do not computerise the data but maintain expensive and cumbersome manual systems. Data collection by some companies is seen as merely meeting the requirements of statutory obligations and few exploit the value of such costly information as a management tool.

Other researchers have found that people are more likely to forget facts and may not be able to recall up to half of what they originally perceived when questioned as little as 24 hours after the event. Gathering good, reliable information must apply to the role of the office manager or indeed any manager and not necessarily for use in disciplinary matters. Whilst the casualty might be excused from recalling certain information concerning an accident because of shock, it is important that procedures are in place which ensure that witnesses are spoken to within 24 hours whenever possible, and that the interview is structured correctly to eliminate some of the problems outlined above.

Published information from the HSE is not solely sufficient to assist you in programme planning. It lacks sufficient detail and in any case, it is not the intention for it to do so. Contributory factors are not identified in sufficient depth to assist with the identification of trends. Once contributory factors are identified then appropriate remedial measures can be introduced using the safety mix referred to in Chapter 1, i.e. by enforcement, education, engineering, environmental strategies or a mix of some or all four. Making something illegal will never prevent it from happening and there will always be a need to combine enforcement with one or more of the other factors or elements within the safety mix.

Because of poor information gathering, large number of safety management decisions are based upon opinion. In order to get good information, then in all accidents one seeks to answer six basic categories of question. These are:

1. Who?
- who was involved in the accident?
- who was the line manager responsible for safety?
- who reported the accident?
- who was called to respond to the accident?
- who was notified?
- who should have been notified?
- who was responsible?

2. When?
- when did the accident occur?
- when were people aware that an accident could occur?
- when did help arrive at the scene of the accident?

3. Why?
- why did the accident happen?
- why were safety practices not applied?
- why did safety procedures fail to work?

4. What?
- what actually happened?
- what were the losses incurred?
- what injuries were sustained?
- what could have been done to avoid the occurrence?

5. Where?
- where did the accident occur?
- where was the safety officer or line manager at the time of the accident?

6. How?
- how did the event occur? For example, suddenly, slowly, without warning?
- how could safety procedures and practices have been improved?
- how does the organisation learn from the accident occurrence?

Further research studies have identified human errors and failings contributing with other factors to the vast majority of accidents and dangerous occurrences. Failing to cope with circumstances leading up to and prevailing at the time of the accident will be evident in most situations. It is important that all accident investigations consider human error and it is therefore vital to identify prime factors involving human behaviour as a major part of the accident investigation, together with human reaction to unsafe conditions within the workplace which should also be included.

There are two primary objectives of accident investigation:

- *accident reduction* – obtaining sufficient data to facilitate the systematic reduction in the type and severity of accidents in the workplace. These will fall into the four categories identified above as the safety mix and will provide information of sufficient quality to balance the mix to best effect
- *accident prevention* – this relates to the application of safety principles in new design and technology whether it is in the area of automation or improvements to the style and type of manufacturing methodology employed in the workplace.

Both of these strategies have to be based upon recognised opportunities which are available whether to the designer, engineer, enforcer, general safety practitioner or office manager for influencing and preventing all forms of accidents from happening. Both strategies require the co-operation of everyone to succeed and must not be left to one person to solve alone. Only in this way can effective strategies be employed.

It is important that the safety representative is involved in all accident investigations and that he/she is fully aware of the aims and objectives of the data collection process. It must be appreciated that in certain circumstances accurate data collection can be hampered by the 'high status' phenomenon and care must be exercised so that the investigation is not seen as a purely disciplinary or enforcement procedure. Such investigations may assist in the formulation of rules and safe procedures but should not be seen as a means to an end. Obvious breaches of established procedures, however, must be dealt with in the interests of everyone concerned.

The most important part of your operational daily safety plan is the systematic collection of accident and dangerous occurrence data for analysis purposes. It should be remembered that many dangerous occurrences happen in reality but never get reported. This may be due to a feeling of failure by the person concerned, and threats of disciplinary action for breaches of safety codes and practice will not encourage the reporting of such incidents. Some organisations have set up confidential telephone lines in an attempt to combat this problem. Some of these firms can show an increase in the reporting of dangerous occurrences but admit that they may not be aware of all such instances. Safety management, therefore, must have a structured and systematic approach towards the gathering of dangerous occurrence data. Whilst the confidential telephone idea is a sound approach to the problem it should not be relied upon as the only method of gathering information. Confidential information forms can also be used and these can be structured in such a way as to collect contributory factors required in the subsequent analysis. A simple example of such a form is given in Figure 4.1. You should have sufficient standing within the organisational structure to be able to interview any member of staff involved in a dangerous occurrence in total confidence. This method is also an important way of obtaining data. It is very likely that an employee involved in such an incident may feel worried by it. He may even feel fright or shock. When a dangerous occurrence takes place it is usually regarded as a good learning vehicle but this practice must not be encouraged! At the same time, if a lesson has been learned by both employee and employer what need is there of disciplinary action?

Having given thought to primary data obtained for decision-making purposes, these should be supplemented wherever possible with other data published by government departments including the HSE, the Department of Social Security, the Government Statistical Office and the RoSPA equivalent abroad. The HSE annually publishes accident and disease types data but does not publish those contributory factors evident in each accident, disease or dangerous occurrence. In fact, the current official data collection form F2508 does not seek such factors contributing to these incidents. It will be necessary, therefore, to collect these locally if the safety mix is to be correctly planned, implemented, monitored and evaluated efficiently and effectively. To provide for the simple evaluation of data collected, it should be gathered in a standardised form. Bearing in mind that the results may be placed upon a computer, thought needs to be given to the coding of the results. To facilitate this, you can break your data collection process into the following broad headings:

- accident details
- casualty information
- details of attendant circumstances.

For administrative purposes, provision will need to be made for the identification of accidents individually and for general recording purposes. Data required by the enforcement agencies need to form a part of this process.

Accident details

In general terms, this section must describe what happened, when the incident occurred and where the location of the accident site was. Within this section it is important that details of all equipment involved at the time of the accident are also recorded. At the same time, accident damage is noted both to equipment and surrounding environment together with details of independent witnesses. It is also important at this stage to note any procedures adopted and whether appropriate safety precautions were taken and that issued safety clothing was worn. Whether procedures were not carried out or clothing not worn is not at issue here. Questions relating to why certain factors exist are dealt with under attendant circumstances and are deemed to be contributory to the accident or occurrence.

ACCIDENT RECORDS | **85**

A. Subject of Report

Fatality ☐	Specified Major Injury ☐	3 Days Injury ☐	Dangerous Occurrence ☐	Flammable Gas Incident ☐	Dangerous Gas Fitting ☐
1	2	3	4	5	6

B. Person or Organisation Making the Report:

Name and Address

Type of Business

Role of Company at Time

Post Code

Main Site Contractor ☐   Sub-Contractor ☐   Other ☐

Name & Telephone Number of Contact

Is Injured Member of Your Family? Yes ☐ No ☐

C. Date, Time and Place of Accident or Incident:

Day Month Year

Time of Accident (24-hour clock)

Location of Accident

Normal Activity Carried Out at Site

D. The Injured Person:

Name and Address

Age ☐   Sex ☐   Employee ☐   Trainee ☐   Self-Employed ☐   YTS ☐   Other ☐

Trade or Occupation

Nature of Injury

Figure 4.1 Sample accident form

THE OFFICE SAFETY HANDBOOK

E. Kind of Accident:

Contact With Moving Machinery ☐	Injured While Lifting/Carrying ☐	Trapped ☐	Exposure to Explosion ☐
Struck by Moving Object ☐	Slip/Trip/Fall Same Level ☐	Drowning ☐	Contact With ☐
Struck by Moving Vehicle ☐	Fall From Height ☐	Exposure ☐	Injured by Animal ☐
Struck Against Fixed Object ☐	Distance Fallen ☐ Metres	Exposure to Fire ☐	Other ☐

F. Agent(s) Involved:

Lifting Equipment ☐	Process Plant ☐	Live Animal ☐	Ladder Scaffolding ☐
Portable Hand Tools ☐	Stored Materials ☐	Moveable Container ☐	Construction Framework ☐
Vehicle ☐	Deficient Atmosphere ☐	Working Surface ☐	Electricity Equipment ☐
Other ☐	Distance Fallen ☐ Metres	Building Mining ☐	Entertainment Sports Equipment ☐

Describe Factors Indicated [_____]   Any other Agent ☐

G. Account of Accident or Dangerous Occurrence:

Describe How the Incident Happened. In the Case of an Accident, State What the Injured Person Was Doing at the Time:

[   ]

Signature of Person Making Report [_____]   Date [_____]

Figure 4.1 (Continued)

Casualty information

Here it is necessary to record details of who was involved in the incident and the levels and type of injuries received. Generally, it is thought necessary to record those killed or injured. Few accident investigators support the classification of injury types at this level. In an accident a person is either killed or injured. A non-injury may be deemed a dangerous occurrence. It is difficult to identify those separate factors which contribute to a serious injury rather than a minor injury or what makes an incident a minor one rather than a serious one. For example, in two identical instances two people fall from a roof 20 feet in the air. One lands on his head and is seriously injured where the other lands on his feet and suffers minor lacerations. In this case, one employee was off work for two months whilst the other returned to work the following day. Where scientific research is involved there are accident injury scales for the classification of disease and injuries. It is argued that for the operational management of effective safety procedures and practices, these are unnecessary. It must be remembered that the aim of the exercise is to identify contributory factors present in when an accident occurs. The sole purpose is to identify these and take remedial action. The fact that an employee was injured in varying degrees does not significantly contribute to this part of the exercise. It is for this reason that you need only classify injuries simply at this stage.

Attendant circumstances

This part of the procedure considers those factors which have contributed to the accident or dangerous occurrence. It will include details of the accident site and its immediate surrounds and will cater for factors which led up to and immediately preceded the accident or occurrence. Where a non-injury incident is involved it may only be necessary for a self-certification exercise by the person involved to be carried out. It is recommended that in these circumstances part 2 of the form, the casualty detail, be suitably modified to ensure anonymity.

Care should be taken that all questions that can be regarded as subjective are completed with care. For example, if an accident occurs outside then it might be necessary to explain the meaning of 'high wind', 'rain/drizzle', 'mist/fog' etc. in any instructions or in training given. These factors often appear in outside-type incidents. Indoor-type accidents can suffer a similar language interpretation if adequate care is not exercised. For example, the difference between bright light, light and darkness will need care as would matters referring to ventilation such as airy, breezy, windy, gusting, etc. In a recent case during a warm sunny spell, the workforce opened windows and doors to let through a cool breeze. The breeze was rather strong and blew over several items of equipment stored in the work area. These fell over and injured a worker's foot whilst another had dust blown into her eyes thus causing the worker to catch her fingers in a machine. Here, the contributory factors are identified as the weather, the open windows and doors, the windy, blustery conditions, the loose materials capable of being blown over, and the dust. The casualty rate could have been affected by the removal of any one (or more) of these factors. Similarly, a person slipped and fell in a car park because it rained for the first time in eight weeks and the rainwater mixed with oil and rubber caused the car park surface to become very slippery. The employee was late for work and in their haste to run into the office to avoid getting wet slipped and broke a leg.

Where someone fails to cope with their environment, most data collected for accident analysis purposes can be standardised. There can, of course, be allowances made for those occasional variations which do exist between occupations and industries.

DATA SOURCES

Good accident investigation techniques require a good data base from which to work. It is necessary systematically to obtain information relating to accidents which will enable you to identify contributory factors. One recognised form of gathering information is by using the survey.

The purpose of surveys is to obtain information about accidents. To obtain this information is never free of cost but can be expensive in terms of time and money. Before collecting any sort of data, thought must be given about how relevant it is to the needs of the situation, how it can be analyzed usefully and how it can be obtained. The following is a simple description of a very complex subject and further reading references are given at the end of the book.

Factors which undoubtedly need consideration when choosing the various sources of data are the:

- worth of the data
- ability to cross check key information
- time required to collect it
- cost of collection
- accessibility (particularly internal records)
- political sensitivity.

Cost benefit aspects should be considered in making the choice. Clearly, subjective judgment will play an important role in this choice as the implication of various data decisions may be unclear.

There is also, a great need to assess the ability to analyze data prior to collecting it. It is not uncommon to discover, after great effort and expenditure of time, that the following are true:

- there is too much data for manual analysis
- the form of the data collected is not united with the form of analysis to be applied.

Some common types of analysis are:

- significance testing of means and variances
  - goodness to fit (X^2) tests
  - testing the correlation between variables
  - establishing a regression line
  - grouping data in various ways
  - preparing histograms, pie charts, graphs, etc.

PRIMARY AND SECONDARY INFORMATION

The establishment of a sound and credible information base is essential for any project evaluation. Common failings in many project and research investigations are:

- a failure to identify potential sources (i.e. insufficient initial research)
- a failure to use sources (being deterred by effort, time, cost or political antagonism)
- poor choice of sources (using subjective data when objective data is possible; failing to build in validation processes; failing to identify the sensitivity of the end decision to data accuracy).

A simple classification of data sources is:

- *primary data* – this is the information which originates from your investigation:
  - what people/systems do (observation and experimentation)
  - what people say (questioning and expert opinion)
- *secondary data* – this information does not originate from the investigation:
  - what people have done/said (internal records)
  - past performance of systems (external published sources).

Primary data

The fact of gathering data for specific needs often forces greater thought about:

- what is trying to be done
- what contribution will the piece of data make?

Gathering data in this way often provides the opportunity for personal contact, obtaining specific pieces of feedback or just developing thought in order to provide a greater insight into the nature of the problem. Almost without exception, some cross checks are needed of secondary data via primary data and vice versa.

Observation

These methods use some kind of recorder, either human or mechanical (e.g. video cameras, or the human eye, etc). It is the aim of such methods to avoid any abnormal behaviour of the system due to the observation. This is far from easy; frequently people have a tendency to settle to 'normal'. Examples of such an approach are:

- activity sampling surveys (working on a new piece of equipment to record what happens)
- filming a specific activity in a department
- traffic flow counters
- visual checks on floor utilisation
- safety studies and research.

Experimentation

The aim is to identify the difference in the outputs of a changed system with those prior to, or in, an unchanged situation. The content of the use is in the testing out of a cause effect model which has resulted from assessment of the current situation. Examples of this might involve the:

- testing of new equipment
- testing new procedures
- testing safety ideas.

Questioning

This is the most common source of primary data involving the use of interviews and/or questionnaires. Asking questions is a good method of establishing and cross checking causal models and is the only way to get at behavioural problems in an organisation. Considerable skill in the use of interviews and questioning techniques is required if data is to be obtained which is worth analyzing as a basis for useful information. The ease with which this source can be misleading or inaccurate is evidenced by the disastrous results of schemes based almost entirely upon data from such sources. The cost and time attached to the use of this source is influenced by the geographic dispersion of the people involved, the number of people involved and the technique chosen. These issues are discussed later.

Expert opinion

There are certain situations where the processing of data to form information is best done by a particular person. Examples might include:

- evidence from a scientist specialising in a particular field relevant to the study
- a production controller estimating delivery of new equipment
- a union representative on reactions of his members to a new procedure.

The reasons for accepting expert opinion in the main fall into the following three categories:

- the rules for processing the data to form information are not available outside the mind of the individual
- the data cannot fully be established as it is in the subconscious as well as the conscious mind
- the worth of extracting the data, etc., is not offset by the benefits of greater accuracy (often a general indication is all that would be required).

Secondary data

Such data tends to be easily accessible and is often a good starting point for an investigation. Desk work based upon such data can be helpful in the following:

- defining terms, gaining understanding in new fields of knowledge
- widening view points on relevant issues
- establishing the current 'state of the art' in specific areas
- identifying the areas where primary data will contribute
- formulating possible approaches to resolving the problem
- obtaining some understanding of the environment under investigation
- providing a theoretical background to the study.

Internal records

Most organisations keep many internal records, some formally, some informally. These records can be of great value in project work once located and released. Skills in interviewing and gaining personal acceptance are often a pre-requisite to this. As a base, internal records can reveal accounts, costings, labour and some accident statistics.

External published data

A large volume of up to date, published data is accessible, some of which is almost bound to be relevant to a specific project. These can sometimes be obtained from professional organisations, trade associations, libraries and colleges of higher education. It is usual for any article to be obtained within 14 days.

Since most problem situations are by no means unique, it is highly likely that there will be some comment in the literature relevant to the problem being investigated. This being so, then it is surprising that the use of external published data is underestimated. However, it is important that when using external data sources the context of any findings is assessed for validity prior to local use.

Comparative studies

The result of collecting primary and secondary data will provide a base for formulating a hypothesis about future behaviour of systems or people which will be significant in the recommendations of any change. The use of comparison and analogy can often give further insight and affect the confidence in a model. The utility of this method lies in the suitability of the comparison chosen.

Comparative studies can offer some of the following:

- a cross check on how to solve a problem
- an indication of similarities and differences in approach
- identification of the results and associated side effects to compare with predictions.

ACCIDENT DATA COLLECTION

Decisions will always be influenced by a person's perception of those elements of the environment, internal or external to the organisation, which are seen as relevant. Thus, a greater understanding of issues and their relevance to a particular decision along with a more realistic picture of the environment will lead to a better basis for decision.

Information is required for many reasons, some of these are to:

- clarify the request of the client
- identify issues relevant to the problem or the decision
- enable projections to be made of future behaviour of people and systems
- establish the current situation and what has led up to it
- formulate an approach to resolving the problem
- check personal ideas against those of others.

There are eight steps involved in establishing a data or information base. These are:

- identifying the data requirements
- an identification of potential sources of data
- a selection of sources to be used
- a selection of the method to be used for collecting data
- planning the method of data collection and the analysis which will follow
- collecting the data
- analyzing the data
- presenting the information in a clear and understandable form.

Data often exists in a form which has little value as an aid to specific decision-making. In converting data into a meaningful form of information, one or more of the following should be considered:

- aggregation or disaggregation of the data
- regrouping and/or ordering of the data
- testing relationships between various types of data
- trying various methods of presenting the data.

It is often only by trying different approaches to, and combinations of the above, that insight and ultimate understanding can be achieved.

What is data?

This is often thought to be numerical consisting only of such things as sets of numbers, graphs, equations and the like. On reflection, it is clear that this cannot be so since many decisions in our daily life are made on the basis of non-numeric information. Data is anything which describes the attributes of a situation, system, person, object, etc. For example, if we wished to describe the safety department or office in which we work, it could be done in a variety of ways:

- number of staff
- sex of staff
- age of staff
- attitudes
- hobbies
- task undertaken
- length of service.

There is almost an infinite amount of data concerned with such a department or office setting:

- some of this data is objective (such as numbers, sex, age, etc.)
- some of the data is subjective (such as attitudes, etc.)

The boundary between objective and subjective is often imprecise and is dependent upon the:

- particular characteristic chosen
- form of measurement used.

How could the safety procedures in your organisation be classified?

Equally, some data are quantitative (i.e. measurable, countable), for example numbers. Some data sets are qualitative (i.e. rankable, assignable), for example hobbies.

Rigorous mathematical methods of anlaysis are available for dealing with both types of data.

Recording data

A specific example used by some organisations to gather data is represented by Stats Form 1 shown in Figure 4.2. This serves to show the level and type of information which could be gathered which is also sympathetic to Form F2508 required by the HSE. Such a form may need modification depending upon the industry in which it is to be used. The important point to remember is that contributory factors need to be identified and any form design should attempt to identify these. The example form shown as Stats 1 would need an accompanying sheet explaining each question.

The first part of the form should give the employer a reference numbering sequence that will identify each individual accident record but also keep a tally on accident numbers per year. It must be remembered that one accident can produce more than one casualty so care must be exercised when carrying out accident analysis as opposed to casualty analysis. On many occasions board members or safety committee members have been confused over accident and casualty figures. Care must be exercised, therefore, when quoting figures.

A note must be kept of the accident location and this can be eased by using factory, building and room numbers for computing purposes. Individual workstations can also be numbered. Locations showing a high accident rate should be investigated separately. It is useful to record accident costs and there are many conventions used to calculate these which are discussed later. However, it is important to keep a record of accident severity and it is only necessary to distinguish between a fatal, injury and non-injury type accidents.

In terms of the accident details then, it is important to record the date, time and exact location of the accident clearly. For road accidents, it will be necessary to use the standard grid reference conventions. Numbers killed and injured should be recorded and whether the accident occurred inside or outside. It will be necessary to list activities carried out by an employee whilst carrying out his/her duties. These should then be classified numerically for computer purposes. Activities or manoeuvres would vary with each industry, therefore careful planning is required. For example, an office worker may have a simple activity chart as follows:

```
00  lifting photocopy machine toner lid
01  securing photocopier lid in the open position
02  travelling to/from photocopying machine location
03  toner accident
04  hot roll burns
05  unpacking new toner
06  removing old toner cartridge
07  replacing toner cartridge in situ
08  cleaning any spillage
09  closing down photocopy machine lid
10  carrying out safety checks on completion
```

Accident type may follow those conventions used by the HSE and as outlined in section E of its Form F2508 as would agents involved and matters referring to flammable items.

ACCIDENT RECORDS | **95**

ABC Manufacturing Ltd

Incident Record Number [ ]   Employee Record Number [ ]

Department [ ]   Location [ ]   Name [ ]

Employee Address [ ]   Post Code [ ]

Incident Class [ ]   Severity [ ]   Man Days Lost [ ]   Cost [ ]

Part 1: Accident Details

Date of Accident [ ] [ ] [ ]   Time of Accident (24 hour) [ ] [ ]   Location [ ]
                Day Month Year

Date of Casualties [ ]   Severity [ ]   Number Killed [ ]   Number Injured [ ]   Damage Only [ ]

Indoor/Outdoor [ ]   Activity [ ]   Accident Type [ ]   Agents Involved [ ]   Flammable [ ]

Part 2: Casualty Details

Sex [ ]   Occupation [ ]   Age [ ]   Status [ ]   Injury Type [ ]   How Long Employed [ ] [ ]
                                                                                Years Months

Qualified [ ]   Training [ ]   Date of Last Course Attended [ ] [ ] [ ]   Course Subject [ ]
                                                              Day Month Year

Course Type [ ]   Previous Accident History [ ]   Date of Last Accident [ ] [ ] [ ]   Previous Accident Type [ ]
                                                                          Day Month Year

Health at Time of Accident [ ]

Part 3: Attendant Circumstances

Light Conditions [ ]   Weather [ ]   Ventilation [ ]   Temperature [ ]   Medications [ ]   Alcohol [ ]

Equipment in Use at Time of Accident [ ]   Date Equipment Last Checked [ ] [ ] [ ]   Checked by Whom [ ]
                                                                        Day Month Year

Equipment Type [ ]   Activity [ ]   Supervision [ ]   Other [ ]

Signature of Person Making Report [ ]   Status [ ]   Date [ ] [ ] [ ]
                                                          Day Month Year

Figure 4.2 Accident and dangerous occurrence reporting, Stats Form 1

Casualty Statement	
Signature	Date

Witness Statement	
Signature	Date

Figure 4.2 (Continued)

Casualty details should be straightforward and should provide information concerning the person injured or involved in a dangerous occurrence. Base data should also include questions relating to the experience and professional competence of the casualty and their ability to carry out their tasks adequately and in safety. Injury type should be simple and might follow the following standardised format:

```
01  head
02  neck
03  shoulder (left)
04  shoulder (right)
05  chest
06  arm (left)
07  arm (right)
08  wrist (left)
09  wrist (right)
10  hand (left)
11  hand (right)
12  fingers (left hand)
13  fingers (right hand)
14  upper torso
15  waist
16  hips
```

17   leg (left)
18   leg (right)
19   ankle (left)
20   ankle (right)
21   foot (left)
22   foot (right)
23   toes (left foot)
24   toes (right foot)

Further codes could be added to identify cuts, fractures, bruising, etc. It may be appropriate to seek medical advice prior to setting up individual injury type codes.

Information concerning a casualty's previous accident involvement is an important consideration as are details of training courses undertaken. Course titles should be listed in a similar manner as those listed above under 'activity'. It has been found necessary by some industries to ascertain an employee's health at the time of the accident and some organisations are able to carry out their own medical examinations periodically. In these cases it is usual to include on the form the date the last company medical examination took place.

Attendant circumstances in the stats form example illustrated in Figure 4.2 lists 'light conditions' as the first question. In some industries adequate lighting is important. In an indoor situation, it might be appropriate to list light conditions as follows:

00   natural light
01   artificial light – standard bulb 60 watts (working)
02   artificial light – standard bulb 100 watts (working)
03   artificial light – standard bulb > 150 watts (working)
04   artificial light – standard bulb 60 watts (not working)
05   artificial light – standard bulb 100 watts (not working)
06   artificial light – standard bulb > 150 watts (not working)
07   artificial light – spotlighting (working)
08   artificial light – spotlighting (not all working)
09   artificial light – neon type (working)
10   artificial light – neon type (not all working)
11   artificial lighting inadequate
12   other (please specify)

Similarly, ventilation as a contributory factor can be itemised. With regards to temperature, internal temperatures should conform to the criteria listed in the Factories Act 1961. For both internal and outside temperatures, the actual temperature as recorded from a thermometer should be used. However, for outdoor accidents, weather conditions may be recorded as follows:

```
01  dry (no wind)
02  dry and windy (Beaufort scales 1 to 12)
03  wet but not raining
04  raining
05  fine rain or drizzle
06  icy (confirmed by temperature reading)
07  snow (but not snowing)
08  snowing (no wind)
09  snowing (with wind) causing drifting
```

Where medications are prescribed to an employee it must be ascertained whether these can affect an employee's performance. Medical advice should be sought in cases where a casualty has taken medications thought to have contributed to the accident. Similarly, alcohol should be monitored. Evidence suggests that alcohol in the workplace is becoming a problem and the only way to discover its involvement in accidents is to seek this information at the outset.

Equipment in use at the time the accident occurred should be itemised together with information concerning the correct use of protective and/or safety clothing. Information regarding equipment/safety clothing maintenance is also important and should be recorded as should details of the person responsible for carrying out these checks.

Equipment in use at the time of the accident should be listed according to those used by the organisation. For example, in the agricultural industry one would need to list (similar to an inventory) all those items of equipment used by employees. These must be coded in a way suitable for computerisation. Within this section it is important to identify whether safety/protective clothing or other measures were used in accordance with laid down instructions.

In some cases, certain items of equipment not deemed to be potentially hazardous may be omitted from the listing until such times as they are involved in an accident. For example, at a local school, a pencil was not regarded as hazardous until one injured a pupil. In this case the pupil had just gone to the front of the class to sharpen his pencil in a desk-mounted drum type sharpener. He had obtained permission from the teacher in charge to do this. On returning to his desk, the pupil tripped over the strap of a school bag and fell forwards over a desk onto another pupil, stabbing the pupil in the arm with the recently sharpened pencil. The lead broke off requiring the injured pupil to undergo hospital treatment. Up until this point, pencils were not a feature of the equipment heading within the computerised data collection process. Since they have been added this particular local authority has experienced several pencil type accidents both in and out of school premises. Remedial action resulted in the removal of the barrel type pencil sharpeners and the teachers now provide small hand held sharpeners to pupils whilst in their seats. This action has significantly reduced the pencil-related type accidents.

Over a period of time, comparisons with the equipment inventory can show those individual items of equipment which feature in accidents against those that do not. This is recognised as a more satisfactory way of establishing which items are dangerous and allows the safety practitioner to build equipment records on a regular basis.

It is important to record casualty supervision details too. All too often office managers and supervisors overlook their responsibilities adequately to supervise subordinates when carrying out their duties. Recent examples concerned the Clapham rail accident and the Zeebrugge ferry disaster. In both cases there was evidence that subordinates were unsupervised. At Clapham a signal repair engineer was said to have left certain aspects of his work in an unsafe condition. His supervisor or line manager failed to notice this omission. Likewise on the Zeebrugge ferry it was claimed that a junior member of staff responsible for closing the bow doors was asleep. His supervisor or line manager failed to notice this and take appropriate action. A general who loses a battle cannot blame his soldiers! With leadership comes a responsibility. This cannot be ignored when things go wrong and those who absolve themselves of this duty show extremely poor management practice. This is particularly noticeable when management blames subordinates, or allows subordinates to take the blame, for not doing their jobs correctly. The process of responsibility must go upwards until it can proceed no further.

It is useful to include on any data collection form a means of obtaining 'other' information which may be regarded as 'special projects'. This can facilitate the recording of information without necessarily requiring a new print run. Although only one box is provided for in the example, some organisations have as many as six 'special project' or 'other' boxes at any one time.

EMPLOYEE CO-OPERATION

Gathering information from employees

This concerns two common methods of collecting primary data; the interview and the questionnaire. There are other methods available such as experimentation, observation, self-recording and so on, but these are relatively sophisticated and rarely applicable to operational project conditions.

The basic decision whether to use the questionnaire or interview is almost always concerned with resources and the scope of the problem:

- how much time and effort and manpower can be allocated to the project?
- where are the sources of data located?
- what kind of data is needed?

The answers to these questions often mean that one of the methods is rejected and thus eases the decision to be made. Other choices can be eliminated by discussing appropriate advantages v. disadvantages whilst in more marginal situations, a choice is made by comparing direct comparison of the two methods.

The questionnaire

Advantages:
- it can be administered to groups of people at the same time thus saving time and expense
- it can be used where the respondents are spread over a wide geographical area
- the required answers on a questionnaire can be pre-structured thus making the task of analyzing much easier
- questionnaires can be more reliable than interviews
- questionnaires can be made anonymous thus allowing for more freedom of response
- questionnaires give the respondent the opportunity to verify factual data
- questionnaires avoid interviewer 'error'.

Disadvantages:
- questionnaires often suffer from a low response rate, which give rise to problems of whether the data collected is representative of the sample population. Respondents can also avoid answering specific questions
- difficult questions cannot be clarified by the respondent
- questionnaires are often seen to be impersonal and respondents do not feel committed to filling them in
- there is no opportunity to ask extra questions on a questionnaire and therefore new data is missed. 'You only get answers to the questions asked.'

The interview

It is important to deal with some general principles of the interview before specific safety examples are discussed. The interview is a face to face verbal exchange, in which one person, the interviewer, attempts to elicit information or expressions of opinion, attitudes or belief from another person or persons.

The role of the interview is to act as a method for collecting data. This may be used:

- during the early stages of an investigation to help identify the problem areas, the relevant dimensions, to suggest hypotheses and to reveal the natural frames of reference existing in the minds of the respondents
- when questionnaires are used, the interview may be employed to pre-test the questionnaire form

- as the main instrument of data collection
- to clarify findings which have emerged from other sources of data.

Advantages:
- interviews increase respondents' commitment
- they tend to be more valid, encouraging true to life answers
- interviews often surface data which would not come out in a questionnaire
- they allow for clarification of difficult questions.

Disadvantages:
- problems of interviewer 'error' such as appearance, manner and style of asking questions
- problems of anonymity or the lack of it
- unstructured interviews often provide data in a form which is difficult to analyze
- interviews are time consuming.

Structured or unstructured?
A structured interview is one in which the questions have been decided upon in advance of the interview. The questions are asked with the same wording and in the same order for each respondent. The essential feature of a structured interview is that the interviewer does not have the freedom to re-word questions, to introduce questions which seem especially applicable to the individual case, or to change the order of topics to conform to the interviewer's spontaneous sequence of ideas. In the unstructured interview, the interviewer technique is completely flexible and can vary from one respondent to the next.

Structured interviews:
- incorporate a basic principle of measurement (that of making information comparable from case to case)
- are more reliable
- minimise errors due to question wording.

Unstructured interviews:
- permit standardisation of meanings, rather than the more superficial aspects of the stimulus situation (the question)
- are likely to be more valid in that they encourage more true to life responses
- are more flexible.

Questions

For both interviews and questionnaires, it is necessary to relate questions to the problems under study. However, the discipline of refining questions and making sure that they are relevant can be difficult. For example, is a question pertaining to an employee's hobbies relevant to the study?

There are a few basic rules that can be applied to the framing of questions:

- avoid ambiguous words and phrases
- avoid long questions
- the questions should state as precisely as possible the time, place and context you want the respondent to assume
- either make explicit all the alternatives the respondent could answer or none of them
- avoid multiple questions
- avoid leading questions
- avoid rhetorical questions
- avoid implied values.

It has been widely assumed that the interview is superior in many ways and must be used whenever resources permit. Certainly the interview should be used at the exploratory stages, particularly in the area of accident investigation. Because of the seriousness of the accident investigation process and the legal requirements to make all workplaces safer such an interview may be referred to as a statement interview or formal interview. This is discussed below.

Statement interview

If statements are considered as part of the operational plan then it is important that they are taken as quickly as possible after the event and in normal circumstances should be taken within 24 hours of the accident happening. As early as 1932, psychologists have been able to show that a person is likely to forget half of what they originally perceived after 24 hours. To take a statement after this period of time can result in memory loss which may affect important detail, or other factors may influence the casualty thus distorting the facts. For example, some research in 1979 by Clifford B, showed that a high status questioner can influence a witness. By this it was possible to show that a motorist stopped by the police would think carefully what he said to the constable because of the fear of prosecution. In this case, the police officer becomes a high status questioner. Likewise, insurance companies receive many reports each year from clients involved in road traffic accidents purporting that trees have mysteriously emerged into the path of their vehicle! In these cases, the insurance company is seen as the high status questioner in that the client's no claims discounts may be involved if they are found blameworthy. In such examples, a distortion of the facts (some refer to this as being economical with the truth) is common.

It should be understood that the taking of statements has a prime objective, and that is to ascertain facts relating to an accident or dangerous occurrence, which will facilitate the introduction of remedial measures which are specifically designed to prevent or reduce the incidence of those events happening in the future. The taking of statements should not be seen solely as a means of apportioning blame or for deciding upon disciplinary action. The safety practitioner conducting such an interview may wish to obtain these facts in total confidence. Some organisations provide confidential phone lines for the reporting of dangerous occurrences. This practice is good and can be recommended.

Disciplinary action must always be used widely and should only be considered after all the evidence has been presented. It is more important that the interview is taken and conducted in a professional manner. These issues are discussed below.

Conducting the statement interview

It is unwise to allow an employee to make a statement concerning an accident without supervision. To do so may mean that vital contributory factors are omitted from the report. It is more appropriate to conduct an interview which is constructed in such a manner as to provide a basis for the systematic gathering of relevant factors pertaining to the accident or occurrence. It is important that these interviews are seen primarily as fact finding exercises.

It has been said that there is a wealth of evidence which questions witness reliability, but also the reliability of the victim statement can be questioned particularly when the consequences of the accident are understood and possible disciplinary or other punitive action could result. Here the interview technique should be used for fact finding only so that the investigator can learn more about the contributory factors which came together to cause the accident. Geiselman and Fisher (1986) describe three types of interview techniques. These are the cognitive, hypnosis and interrogative type interview. Here, hypnosis means what it says and the interrogative type interview is that which one would expect from the police. The most effective methods found by these two authors were the cognitive and hypnosis methods. the latter was rejected for practical terms due to the length of training necessary to hypnotise people. The ethics of this are also questionable. The cognitive type interview was found to be equally as good and easier to learn. Basically, this method consists of four general points for jogging memory plus several specific techniques which are outlined below. There is no reason why these points could not assist you to obtain more reliable information from injured members of your workforce. The four points are explained to the interviewee before the narrative report. The first two attempt to increase the overlap of elements between the stored memory and the retrieval mechanisms. The last two encourage the use of several retrieval paths. The four points are:

Reconstruct the circumstances
Here the interviewer asks the witness or victim to reconstruct the incident in general terms. Reconstruct the circumstances that surrounded the incident. Get the interviewee to:

- think about the environment in which the accident happened, such as room layout, furniture locations, weather, vehicles, equipment, lighting and other people or objects
- think about what they were doing before the accident happened, how they were feeling and their reactions to the incident.

It is important to start before the accident happened and it may be necessary to start 24 hours before the accident happened as evidence exists to show that stress at home, alcohol and other factors can feature as major contributory factors in an accident happening.

Report everything
The investigator should explain that all information may be helpful and nothing which is thought to be irrelevant should be omitted. All information should be given. Some people tend to think that some information is not important so they hold back. Do not ask your casualty or witness to edit anything.

Recall events in a different order
It is natural to ask the person being interviewed to start at the beginning and journey through to the end. If you have done this, it is worthwhile asking for the information in reverse order or try starting with the point that concerned the interviewee the most about the event. From here you may go forward or backwards in time.

Change perspectives
Try to see the event from another person's point of view or adopt the perspectives of others who were present at the scene.

Mental reconstruction of circumstances surrounding a memorable event has been shown to be a powerful aid to memory and visit to the accident site and undertaking accident reconstruction exercises are valuable.

In addition to the four points outlined above, the cognitive interview also uses a series of specific questions to help an accident investigator elicit information following the narrative phase of an interview. The investigator might consider the following:

- was the casualty qualified to carry out the task being undertaken at the time of the accident?
- was safety clothing worn/safety equipment used?
- when did the casualty last undertake relevant safety training?

- when was the equipment involved in the accident last examined?
- were safety procedures followed?
- was the casualty fit to undertake his duties?
- have the contributory factors been identified?
- how could these have been eliminated and how can they be for the future?

These questions are of a general nature and should not be regarded as exhaustive. They are given as a general indication of the type of information which must come out of an interview situation when accidents have occurred. The aim of the exercise must be accident prevention in the future and a decision will have to be taken as to the blend of safety mix required. We know that enforcement on its own is ineffective so taking disciplinary action alone could not be recommended. It must be used with other measures which must be elicited from the investigation.

ACTION CHECKLIST

- *Do you understand the importance of accident investigation as a tool for accident reduction programme planning?*
- *Do you have an adequate accident investigation procedure?*
- *Are your data collection and analysis procedures adequate?*
- *Do you have an effective means of collecting dangerous occurrence information?*
- *Have you a procedure for interviewing witnesses for all levels and type of accident and/or dangerous occurrence?*
- *Do you interview accident casualties with the sole purpose of identifying contributory factors to the incident?*
- *Are the procedures for storing this information appropriate and adequate for the purpose?*

'... COSHH Regulations introduced a new legal framework for the control of substances hazardous to health ...'

'Make sure your workforce is trained and instructed in the risks and precautions to take ...'

What to look for in the office

FIRE

It is necessary to consider the following:
- fire precautions legislation
- fire appliances and equipment
- fire signing and escapes
- fire certificates
- fire insurance
- fire prevention
- fire procedures

Fire precautions and legislation

The relevant law concerning fire precautions are to be found in the following:
- Fire Precautions Act 1971 (and regulations and orders made thereunder)
- Fire Safety and Safety of Places of Sport Act 1987
- Health and Safety at Work Act 1974
- The Building Act, 1984
- Fire Service Act 1947
- Building Regulations 1991 (SI 1991 No 2768)
- The Fire Precautions (Factories, Offices, Shops and Railway Premises) Order 1989 (No 76)

The prime piece of legislation is the Fire Precautions Act 1971 which was amended by the Fire Safety and Safety of Places of Sport Act 1987. This act applies to all premises in use irrespective of whether they are industrial, commercial or public. The local fire authority is empowered to enforce these requirements under duties and powers specified in the Fire Services Act of 1947. Generally, all fire provisions in industry and commerce in respect of fire escapes, alarms and the like are the responsibility of the fire authority but premises containing hazardous, flammable and/or explosives are controlled by the HSE. Section 2 of the HSWA 1974 provides for fire orientated measures to be undertaken by employers in respect of fire extinguishers, fire training and drills.

Fire appliances and equipment

There are basically five types of fire extinguisher:

- water extinguishers coloured red
- foam extinguishers coloured cream
- dry powder extinguishers coloured blue
- carbon dioxide extinguishers coloured black
- vaporising extinguishers coloured green

What each is used for is shown in Table 5.1.

Table 5.1 Fire appliances and equipment checklist

Class of fire	Description	Extinguisher Type
A	Solid materials usually organic with glowing embers	Water, foam, dry powder, varporising liquid or CO_2
B	Liquids and liquifiable solids such as: those miscible with water, acetone or methanol, those immiscible with water such as petrol benezene, waxes and fats	Water, foam (ensure estability in miscible solvents), dry powder and CO_2 Foam, dry powder, CO_2 or vaporising liquid

Class of Extinguisher	Colour	Quantity	Location	Date Last Checked
Water	Red			
Foam	Cream			
Carbon dioxide (CO_2)	Black			
Dry powder	Blue			
Vaporising liquid	Green			

Kitchen areas should also be provided with a fire blanket.

A manually operated fire alarm is required in all premises required to hold a fire certificate (see below). Such a system must comply with the current code of practice for the installation of fire alarms as should appropriate fire equipment used. Such a system should be tested every three months and the results recorded. Emergency lighting may also be required.

It may be necessary to provide fire wardens (see below) with a torch and an armband so that they are easily recognisable during a building evacuation.

Fire signing

It is important that all escape routes are adequately signed and that such signing is kept clear and can easily be seen. Damaged or old, dirty signs should be replaced regularly. All escape routes must be kept clear and unobstructed and should afford for quick and easy evacuation of a room or building. All means of escape should provide the following:

- all doors affording a means of escape though not in continuous use should be clearly indicated
- sliding doors should clearly show the direction of opening
- doors affording a means of escape should be regularly maintained
- doors affording escape should not be obstructed or locked.

No smoking areas should be clearly identified and staff made aware of such rules and regulations. Combustible waste should be kept away from smoking areas and in metal containers with appropriately tight fitting lids.

Smoking should only be permitted in well ventilated areas aware from combustible or flammable liquids materials. Ashtrays must be provided with a sand or water base and must be used.

Where a building has several floors serviced by a lift then each floor should be clearly marked so that visitors or new employees can readily identify where they are and follow the appropriate fire exit signs. Stairways should be similarly marked.

Fire certificates

These are issued by:

- local fire authorities
- the Health and Safety Executive.

It should be noted that the situations requiring compulsory certification under the Fire Precautions Act 1971 have been deregulated under the Fire Safety and Safety of Places of Sport Act 1987. A certificate may be required in the following circumstances:

- buildings containing two or more factory and/or office premises where the number of persons employed in all of them at any one time is in excess of 20 people
- buildings containing two or more office and/or factory premises where the number of persons employed at any one time in all of them other than on the ground floor exceeds 10 persons

- factories where explosives or highly flammable materials are stored or used in or around the premises unless in the opinion of the fire authority there is no serious risk to persons in the vicinity.

A fire authority may grant an exemption from the certification requirements of the Fire Precautions Act 1971 as amended under the Fire Safety and Safety of Places of Sport Act 1987. If you are in doubt as to your particular requirements you should contact your local fire prevention officer for further advice.

Fire insurance

The general requirements which apply to insurance are covered in Chapter 3. However, there are some specific points to consider which apply specifically to fire insurance. It is usual for only direct loss caused by fire to be insured. It should be noted that damage to property caused by falling debris will be insured if destroyed by fire but damage caused by subsequent explosion would not. It is important that all insurance policies are reviewed annually to ensure that adequate cover is provided. If in doubt contact your insurance broker.

Fire prevention

Good housekeeping, safe storage of flammable materials and appropriate no smoking rules are usually cheap and simple ways of reducing the risk of fires. For organisations using or storing highly flammable or explosive materials, you must provide storage areas designed especially for this purpose. If you have any doubts as to your duties and responsibilities in this respect you should seek professional advice regarding the appropriate steps to be taken. The local health and safety inspectorate or local fire authority would be delighted to assist in such cases. This advice is usually free of charge.

Fire training is an important aspect of fire prevention and it is important that key personnel are familiar with basic fire fighting techniques. Action which is taken quickly and correctly can prevent a fire spreading. There are three fundamental aspects to extinguishing a fire:

- *starvation:* this can be achieved by removing the fuel from the fire, isolating the fire from the fuel source or reducing the bulk of the fuel source present
- *smothering:* a fire also requires oxygen to burn. By removing the supply of oxygen by covering the fire with a fire blanket or an inert gas
- *cooling:* dowsing the fire and fuel with water.

Table 5.2 illustrates the types of fire and extinguishers to be used. The fire brigade should be called whenever a fire is detected, however small you think it might be.

Table 5.2   Types of fire extinguisher to be used

Type of Extinguisher	Colour	Fire Type
Water	Red	Suitable for ordinary fires involving wood or paper but not flammable liquids. Not to be used for fires involving electricity
Foam	Cream	Suitable for small liquid fires or small oil fires where foam can be used to smother the fire. Foam may not put out a fire on a vertical plane
Dry powder	Blue	Suitable for flammable liquids and electrical fires
Vaporising	Green	Suitable for electrical or electronic equipment fires. May be used for flammable liquid fires bu does produce toxic gases. The hotter the fire the worse are the toxic gases
Carbon dioxide	Black	Suitable for fires involving electrical equipment

Fire procedures

It is important that fire procedures are communicated to all staff regularly. Such procedures should be tested at regular intervals in order to evaluate their effectiveness. Each department or floor should have a fire warden appointed who is responsible for ensuring that all persons are evacuated safely in the event of a fire or fire drill. Such a person should be in possession of a torch and a record of everyone working in the area of their responsibility. This must include visitors. To facilitate this it is necessary for all employees to sign in or out. When the alarm is sounded, all employees and visitors will proceed by the most direct route to their preselected assembly points. The fire warden will immediately search rest rooms, toilets and other places in order to ensure that everyone is evacuating the premises. On arrival at the assembly point a role call must be taken. The fire incident officer may require this information in the event of a real fire.

You must ensure that this procedure is communicated to all staff and that it is regularly tested. Everyone must book in or out and locations must be known by the fire warden. In the event of a real fire, it is totally unacceptable to ask fire officers to enter dangerous buildings looking for missing members of staff only to discover they were not on the premises anyway.

It is necessary to publicise all fire procedures and assembly points.

FIRST-AID

The obligation to provide first-aid facilities comes within the general duties under the HSWA. This requires employers to ensure a healthy and safe working environment for their employees. Section 2(1) requires that employers will ensure, so far as is reasonably practicable (see Chapter 2), the health, safety and welfare of all their employees. This extends to the provision of appropriate first-aid facilities. The Health and Safety (First-Aid) Regulations 1981 provide for the following:

- provision of first-aid equipment and facilities
- first-aiders
- first-aid room where required
- first-aid training.

Equipment

An employer must provide adequate equipment and facilities to enable first-aid to be rendered to employees who:

- are injured at work
- become ill at work.

There are four criteria used to decide what provision is necessary:

- the number of employees
- the nature of the business
- the size of the organisation and spread of employees
- location of the business and employees place of work

A summary of the criteria is given in Table 5.3.

Training

The first-aid regulations require that an employer must provide or make available an adequate number of suitable people who have been trained in first-aid. A person is not regarded as suitable unless that person has undertaken a course of first-aid training which has:

- been approved by the HSE, or
- been provided with additional specialised training so approved.

A suitable person may be:

- a first-aider
- an occupational first-aider
- or other person who has approved training and qualifications (e.g. Registered General Nurse or medical practitioner).

Table 5.3 First-aid cover requirements

First-aiders	Employees
1	50
2	150

First-aid room	Employees
1	250

Where premises are said to be hazardous such as shipbuilding, factories, warehouses and farms there should be a trained first-aider as outlined above at all times. Where a first-aider is away or off work, cover for the absence must be provided by the employer. Where an organisation employs fewer than 50 persons then an employer must still provide for first-aid cover.

A first-aider is a person who has received approved training and holds a valid first-aid certificate.

First-aid training

It is essential that there are sufficient qualified first-aiders available on duty at any one time so as to allow for absences. Unless an organisation has sufficient numbers to train, it would probably not be financially viable to run an in-house course. However, if such a localised demand exists, then the course content, trainers and testing facilities will need to be approved by the HSE. It is usual for the smaller company to contact:

- the St John's Ambulance Brigade
- the British Red Cross
- the Health Promotion Manager of your local Health Authority

All of these organisations will be able to advise on appropriate approved certification first-aid courses.

First-aid rooms

Such a facility is required where 400 or more employees are at work. It is expected that such a facility will be appropriately staffed. Appropriate staff requires a minimum qualification of an occupational first-aider where:

- there are establishments with special hazards
- the place of work is a construction site with more than 250 persons at work
- access to casualty centres is difficult.

Otherwise, a certified first-aider is necessary.

Access to a first-aid room should be available at all times when at work. Such a room should be sited in an appropriate location so as to facilitate vehicular access so that injured persons may be transported to hospital with the minimum of inconvenience to everyone, including the workforce. Such a room should:

- contain suitable first-aid equipment and facilities
- be properly ventilated
- be adequately heated
- have sufficient lighting
- be clean and tidy
- be properly maintained
- be large enough to house a couch
- have suitable access and egress for a stretcher, wheelchair or carrying chair
- indicate clearly the names and locations of the nearest first-aiders

First-aid boxes should be the responsibility of the qualified first-aider. Do not leave first-aid boxes open and unchecked otherwise when they come to be used they will probably be empty!

Self-employed personnel are required to provide themselves with adequate first-aid facilities and as such should carry their own first-aid kit. Professional drivers who are employees or self-employed should also be provided with/or provide for themselves (if self-employed) an approved first-aid kit for use in their vehicle whilst away from their depot or works.

HEATING

The legal requirements covering the provisions in respect of temperature in places of work is found in section 3 of the Factories Act 1961 and in the Offices, Shops and Railway Premises Act 1963. Also, within the general terms of the HSWA every employer is required to provide and maintain a safe and healthy working environment. This will include temperature and humidity. A thermometer must be provided and the minimum acceptable temperature in an environment where workers are expected to sit for a great proportion of their time is 16°C. This temperature must be reached and maintained within one hour from the first hour of work.

There are specific regulations and statutory obligations relating to different industries such as the textile industry or where artificial humidity is produced. A general summary is provided in Table 5.4.

Table 5.4 Some common temperature requirements

Workplace	Temperature	Essential requirements
Office	16°C (min)	Thermometer
Factory	16°C (min)	Thermometer
Steam room	22.5°C (max)	Hygrometer
Other room where artificial humidity is produced	22.5°C (max)	Hygrometer

LIGHTING

Most modern buildings will meet the legal requirements regarding the provision of good adequate lighting. The requirements to provide good and adequate lighting can be found in the Factories Act 1961, the Offices, Shops and Railway Premises Act 1963 with further requirements to provide a healthy and safe working environment within the terms of the Health and Safety at Work Act 1974. If an employer fails to provide and maintain a suitable standard of lighting he will be in breach of section 5(1) of the Factories Act 1961. It is important you set up a system whereby lighting (e.g. bulb replacement) is monitored regularly and that light bulbs provide sufficient lumens per watt where natural lighting is deficient. It is not the intention here to produce a lighting engineer, but to provide sufficient information to provide monitoring information. Table 5.5 summarises all you need know for daily operational purposes.

Table 5.5 A standard lighting checksheet

Type of Lamp	Lumens per Watt	Location	Number	Date Last Checked
Incandescent	10 to 18			
Tungsten halogen	22			
High pressure mercury	25 to 55			
Tabular fluorescent	30 to 80			
Mercury halide	60 to 80			
High pressure sodium	100			

(Note: Incandescent lamps are the common coiled filament lamps.)

Further advice, where necessary, should be sought from an appropriately qualified lighting engineer.

VENTILATION

Section 2(1) of the HSWA is regarded to include a duty to provide all workers with an adequate supply of pollution-free or uncontaminated air. Because employers must provide and maintain a safe and healthy working environment, they may be criminally liable if the workplace is not adequately ventilated and dust or contaminant free. Moreover, in section 2(2) of the HSWA all employers must inform, instruct and train their staff in health and safety procedures. This means that employees must know how to use, test and maintain equipment that ensures a pure air supply and controls air pollutants such as dust and fumes. In addition, sections 4 and 63 of the Factories Act 1961 contain two separate provisions in respect of:

- air purification (section 4)
- dust and fumes (section 63).

Air purification

A circulation of fresh air is necessary to:

- ensure and maintain the adequate ventilation of places of work
- render harmless, as far as possible, all fumes, dust and other impurities which may be harmful to health or are generated in the course of any process carried out in the factory.

Where section 4 of the Factories Act 1961 might exclude certain areas such as boiler-houses or the provision of breathing apparatus, the HSWA requires an employer to take all reasonable steps to ensure the health and safety of the workforce.

Controlling dust and fumes

This applies where:

- any dust or fumes or other impurities which are likely to impair health or are regarded as offensive to the employee
- substantial quantities of dust of any kind is given off.

Employers must take measures to:

- protect employees against the inhalation of dust, fumes and/or other impurities
- prevent dust from accumulating in the workplace
- provide a free flow of clean air.

A list of specific regulations referring to dust and fumes is given in Table 5.6. Recognised methods of controlling airborne pollution is given in Table 5.7.

Table 5.6 Some regulations covering dust and fumes

Regulation	Relevance YES	Relevance NO
Control of Asbestos at Work Regulations 1987		
Grinding of Metals (Miscellaneous) Regulations 1925 and 1950		
Grinding of Cutlery and Edge Tools Regulations 1925 and 1950		
Blasting (Casting and Other Articles), Special Regulations 1994		
Foundries (Parting Materials) Special Regulation 1950		
Non-Ferrous Metals (Melting and Foundings) Regulations 1962		
Chemical Works Regulations 1922		
Indiarubber Regulations 1922		
Chromium Plating Regulations 1931		
Iron and Steel Foundries Regulations 1953		
Highly Flammable Liquids and Liquified Petroleum Gases Regulations 1972		
Control of Lead at Work Regulations 1980		
Factories (Flax and Tow Spinning and Weaving) Regulations 1906		
Factories (Flax and Tow Spinning and Weaving) Regulations 1907		
Jute (Safety, Health and Welfare) Regulations 1948		
Construction (General Provisions) Regulations 1961		
Smoke Control Areas Regulations 1990		
(Continue this list and keep it up to date Also, refer to Table 2.3)		

Table 5.7 Some ways of controlling airborne pollution

Type	Hazardous Substance	Source	Process	Possible effect	Possible Remedi
DUST	Silica (Clay) Hardwood Spore	Dry sweeping Sanding Mould hay	Pottery Woodwork Agriculture	Silicosis Nasal cancer Famrer's lung	Dampen down Respirator Ventilation
FUMES	Zinc Cadmium	Hot flame Heat	Flamecutting Hard solder	Fume fever Emphysemia	Ventilation Respirator Extractor
GAS	Nitrogen oxides Carbon dioxide Slurry gases	Hot flames Engine exhaust Fermentation	Welding Garages Agriculture	Lung irritation De-oxygenation Asphyxiation	Ventilation Respirator Extractor
VAPOUR	Perchloroethylene Isocyanate	Evaporation Moulding	Dry cleaning Plastics	Liver damage Asthma	Ventilation Respirator Extractor
MIST	Chromic acid Non-solvent oil Animal infection	Bubbles breaking Machine lubricant Meat handling	Plating Engineering Abattoir	Ulceration Skin cancer Brucellosis	Ventilation Protective Clothing Respirator

NOISE AND VIBRATION

There is a requirement under the HSWA to provide a healthy and safe working environment. It has long been accepted that exposure to severe noise and vibration can seriously threaten a person's health. In addition, the Noise at Work Regulations 1989 require:

- employers to prevent damage to hearing
- designers, manufacturers, importers and suppliers to prevent damage to hearing.

Employers are deemed to include self-employed persons and an employee must co-operate with their employers' programme to prevent hearing damage. Within these regulations, every employer shall ensure that a competent person makes a noise assessment which is capable of:

- identifying which employees are exposed
- providing information with regard to noise exposure levels
- providing a review procedure of noise levels where necessary.

Noise assessment records should be kept so that trends may be identified over longer periods of time. A sample noise assessment form is given in Table 5.8.

Every employer shall undertake a programme of measures to:

- identify noise sources
- identify remedial measures to be taken
- implement those remedial measures
- ensure that action is taken
- monitor the situation
- reassess noise exposure levels.

Factors to consider are:

- the number of workers who would benefit by the noise reduction programme
- the noise exposure levels involved
- a socio-technical appraisal of noise reduction strategies
- factors which might impede wearing of ear defenders.

Table 5.8 A sample noise assessment form

Location	Background noise level	Peak noise level	Date of assessment	Action taken	Person responsible

In areas where noise levels require remedial action, appropriate signing must be displayed warning that ear protectors must be worn.

There is a need to take action if workers receive a daily personal exposure to noise at or above 85dB(A) or at a peak sound pressure at or above 140dB(A)

A checklist is provided in Table 5.9.

RADIATION-VDUs

In Table 5.10 there is a list which illustrates the various types of electromagnetic radiation and shows some of the hazards that they can cause. The most common exposure to the modern office is the low frequency radiation emitted from the computer visual display unit (VDU). Some problems connected with prolonged use are:

Table 5.9 A sample noise level checklist

Noise level	Checklist	YES	NO
>85 dB(A)	Are noisy machines or processes identified by warning signs?		
	Does everyone in the noisy area need to work there?		
	How long can people stay in the noisy area?		
	Have employees been warned about the dangers of noise?		
	Have they ear protectors?		
	Do they wear ear protectors?		
	Has manufacturer's information about noise levels been checked?		
	Will changes in work methods affect noise levels?		
	Can noise be reduced by fixing loose/vibrating pieces?		
	Can better maintenance reduce noise levels?		

Comments:

Action taken:

Signed: Date:

WHAT TO LOOK FOR IN THE OFFICE | 121

- complaints of eye discomfort such as burning, itching and soreness
- complaints concerning blurred vision, focusing problems and squinting
- posture complaints such as headaches, backaches, tiredness or sleepiness.

Workers exposed to prolonged VDU use should be given regular eyesight checks and steps should be taken to monitor:

- lighting
- location of VDU
- type of workstation used
- seating
- contrast and brightness controls
- length of time spent at the VDU
- size and type of VDU used
- appropriate screen filter.

Workers exposed to ionising radiation may need to:

- undergo regular medical examinations
- undergo regular dose assessments
- appoint a specialist radiation protection adviser
- have contingency plans to cater for radioactive spills etc.
- obtain authorization for the use, storage and disposal of radioactive waste.

PROTECTIVE CLOTHING AND EQUIPMENT

Statutory requirements for the provision of protective clothing are laid down by several regulations made under the Factories Act 1961. Although no specific requirement is made under the HSWA regarding protective clothing/equipment it does state that employers cannot charge for providing it. Not only can he not charge for it, such equipment must be readily available and accessible for immediate use. Where such equipment or clothing is provided, an employer must also ensure that it is used or worn correctly.

Within the general terms at common law, the duty of care which every employer owes his workforce requires that he must protect his employees from the risk of reasonably foreseeable injury. Whether a hazard is regarded as reasonably foreseeable must depend upon the circumstances and whether this type of incident has been experienced before.

In Table 5.10, a summary is provided which contains the requirements for protective clothing in some well-known industries and illustrates which regulations refer to each requirement.

Table 5.10 Example list of protection requirements

Application	Required Protection	Regulations
Workplaces where bottles and syphons are manufactured	Aprons Face guards Footwear Gauntlets	Aereated Water Regulations 1921
Workplaces where asbestos is used	Breathing apparatus Head covering Footwear Overalls	Control of Asbestos at Work Regulations 1987
Workplaces where construction and works of an engineering nature carried out	Eye protectors Helmets Gloves Clothing Belts	Construction Regulations 1961 and 1966
List those carried out by your organisation	*List those items of protection you must provide*	*List those regulations which relate to your industry*

It is important that safety equipment and clothing conform to the appropriate safety standards and these are published by the HSE. Details are provided in the supplementary reading list at the end of this chapter. Standard personal protection items would include:

- safety eye glasses
- footwear
- hats
- face shields
- gloves
- leggings
- waterproofs
- ear protection
- neck protection
- respiratory equipment.
- VDU screen filters.

If you provide such items you must ensure that they are worn and or used correctly at all times.

MANUAL HANDLING

The incorrect handling of products and materials has been identified as a contributory factor in the majority of minor accidents and minor injuries in the workplace. Factors to consider are:

- weight of the load
- shape and size of the load
- height of the load
- material the load is made of
- condition of the floor
- material in which the load is contained
- available headroom
- space around the load
- temperature of the room in which the load is stored
- lighting conditions.

Many training organisations now teach kinetic handling techniques which improve balance and posture for the manual aspects of load moving. This is particularly important when using ladders (where files are stored on high shelving or when lifting heavy files from awkward positions).

Transport

If your organisation operates transport then make sure that someone is responsible for it. In this way vehicle maintenance, driver training and so on can be effectively delegated. It is important to ensure that:

- drivers are properly trained
- where appropriate drivers are correctly licensed to drive the vehicle on the public highway
- only authorised drivers use the vehicles
- vehicle faults are rectified as quickly as possible
- visitors are aware of transport rules and procedures
- supervision is available when undertaking certain manoeuvres
- trackways/roadways are properly maintained and unobstructed
- pedestrians and vehicles are safely segregated
- loading bays are away from general traffic
- drivers are protected from falling objects or vehicle roll over.

Where vehicles have to be repaired make sure that the repairs are carried out by a competent person and that such repairs are carried out in an area specifically set aside for the purpose.

Stacking and storing

Not only is poor stacking responsible for many injuries each year, but for damaging the stored items themselves. Ensure that:

- items which can roll are properly checked
- pallets are not damaged
- containers are serviceable
- racks are not damaged
- stacked pallets are vertical on a level floor without overbalancing
- packages are stacked in brick fashion so that there are no independent columns
- where racks are used they are fixed to the wall
- items do not block gangways/walkways or protrude
- ladders are used at all times for climbing racks
- heavy stacks are not leaning against retaining walls
- de-stacking is from the top only
- safe loading limits of racks are not exceeded.

Lifting equipment

Where lifting equipment is used ensure that:

- test certificates for all lifting machinery is up to date
- safe working loads are observed by all operators
- annual or six-monthly examination reports are available
- only certified lifting equipment is used
- safe working loads are not exceeded
- loads of doubtful weight or machine adequacy are not used
- load centres of gravity are known
- damaged or makeshift equipment is not used
- chains, wires, ropes are not damaged or worn
- appropriate padding is available to avoid damage to the load when lifted
- loads are lifted slowly and not snatched
- cranes have the correct counterweight, load radius indicator and/or safe load indicator
- a responsible slinger or banksman uses a recognised signalling system
- loads do not drop from great heights
- access to motor rooms and service pits of hoists and lifts are correct.

COSHH

The Control of Substances Hazardous to Health (COSHH) Regulations introduce a new legal framework for the control of substances hazardous to health in all types of businesses, including factories, farms, quarries, leisure and service activities, offices and shops. The regulations require you to make an assessment of all work which is liable to expose any employee to hazardous solids, liquids, dusts, fumes, vapours, gases or micro-organisms. Assessment means evaluating the risks to health and then deciding on the action needed to remove or reduce those risks.

The responsibility to make the assessment rests with you – the employer. As the employer, you could do or lead the assessment yourself or give the task to someone else with the authority and ability to get all the necessary information and make correct decisions about the risks and the precautions that are needed. That person should know the point of the various requirements of the COSHH Regulations and have access to a copy of the regulations and approved code of practice. Whoever does the assessment, you should make sure that managers, supervisors and employees' safety representatives are fully consulted about the work processes, about what workers are doing (or are liable to be doing), and about the risks and the necessary precautions.

In some cases, particularly if you are in doubt over the answers to the next questions, you may need to consult your supplier or trade association or even obtain expert advice about what substances are involved in the work. Ask yourself if employees are liable to be exposed to hazardous substances in your workplace. Include service activities as well as production processes.

How can substances hazardous to health be identified?
- For substances brought in, check the safety information on the labels and the information for safe use provided by your suppliers (they are required by law to do so).
- Use your existing knowledge (eg past experience, knowledge of the process, understanding of current best practice in your industry, information on work related health problems in your industry).
- Ask your trade association and other employers in the same business for their experience and advice.
- Check COSHH: is the substance mentioned in any of the regulations or schedules? Is it listed in HSE Guidance Note EH40?
- Examine published documentation, trade data, HSE guidance material.
- Check Part IA1 of the approved list issued under the Classification, Packaging and Labelling of Dangerous Substances Regulations 1984: anything listed as very toxic, toxic, corrosive, harmful or irritant comes under COSHH.

Do the ways in which each substance is handled or is present in the workplace give rise to any risks to health in practice now or in the future?

Question	Corrosives	Acids	Solvents
What is brought into the workplace?			
What is used, worked or stored?			
What substances are produced at the end of the work process?			

	Dust	Fumes	Gases	Residues
What is given off during any process or work activity?				

	YES	NO
Has the HSC approved an occupational exposure standard for the substances and is it listed in Guidance Note EH40/89?		
Is it listed in Part 1A of 'Information Approved for the Classification, Packaging and Labelling of Dangerous Substances' as being very toxic, toxic, harmful, corrosive or irritant?		
Is the substance present at a substantial concentration in the air?		
Is there a minimum exposure limit to the substance?		
Is it a micro-organism which can cause illness?		

(**NB:** If the answer is 'YES' to any of these questions, it may be a substance hazardous to health within the meaning of the COSHH Regulations. It depends on whether it arises out of or in connection with work under the employer's control.)

Figure 5.1 A typical COSHH preliminary checklist

Observe, find out about and consider

- Where and in what circumstances are substances used, handled, generated, released, etc? What happens to them in use?
- Is their form changed (e.g. solids reduced to dust by machining)? Identify places (e.g. handling departments, storage areas, transport).
- What are people doing; what might they do?
- What measures are currently taken to control exposure and to check on the effectiveness and use of those measures?
- Who will be affected (e.g. employees, employers, contractors, public)?
- Is exposure liable to occur?
- Is it likely some of the substance will be breathed in?
- Is it likely to be swallowed following contamination of fingers and/or clothing?
- Is it likely to cause skin contamination or be absorbed through the skin?
- Is it reasonably foreseeable that an accidental leakage, spill or discharge could occur (e.g. through breakdowns of the plant or control measures or operators' mistakes)?

Reach conclusions about people's exposure: who, under what circumstances, the length of time they are or could be exposed for, the amount they are exposed to, and how likely exposure is to occur. Combine this with knowledge about the potential of the substance for causing harm (i.e. its hazard) to reach conclusions about the risks from exposure.

Sometimes, of course, the quantities, the exposure time or the effects are such that the substances do not or could not constitute a risk – but you must have the information to back up this conclusion.

Action to be taken

If the assessment shows that there is no likelihood of a risk to health, the assessment is complete and no further precautions are needed.

If the assessment shows that further action is needed, you have to decide what needs to be done to complete the assessment requirements. If it is reasonably practicable to do so, you should prevent anyone from being exposed to any hazardous substances. Where it is not reasonably practicable to prevent people being exposed, you have to ensure their exposure is adequately controlled and their health protected. In such cases you will need to:

- select the measures to achieve and sustain adequate control
- work out arrangements to make sure those control measures are properly used and maintained
- make sure your workforce is trained and instructed in the risks and the precautions to take, so that they can work safely. In some circumstances, employees' needs to be monitored and arrangements made for them to be under health surveillance (check COSHH, HSE guidance notes relevant to your work and trade literature).

Unless you can easily report and explain your conclusions at any time because the assessment is simple and obvious, you should make a record of it. Record or attach sufficient information to show why decisions about risks and precautions have been arrived at and to make it clear to employees, engineers, managers, etc. what parts they have to play in the precautions.

If the conclusions alter, for example, the introduction of a new process or machine or a change in the substances used, or if there is any reason to suspect that the assessment is no longer correct, for example, reports of ill health related to work activities, the assessment must be reviewed to take account of these new circumstances.

WELFARE

Within the general terms of welfare it is necessary to provide sufficient and hygienic toilet and washing facilities in all places of work. There is also the requirement to provide washing facilities where the work is particularly dirty or arduous. These requirements will be found under the Factories Act 1961, the Offices, Shops and Railway Premises Act 1963 and the statutory instruments made under them. In addition, the HSWA also requires that a safe and healthy working environment be provided and that adequate facilities and arrangements must be made for employees' welfare. This will include the provision of adequate and hygienic toilet and washing facilities.

All sanitary conveniences provided by an employer must be kept clean and there must be provision for lighting, ventilation and privacy. Where persons are employed of different sex then proper and separate accommodation

Table 5.11 A summary of the Sanitary Accommodation Regulations 1938 (Amended 1974)

Minimum number of conveniences	
	Toilets
For every 25 female employees	1
For every 25 male employees	1
Where the number of males exceeds 100 and sufficient urinal facilities are also provided	4
For every 40 employees over 100	1
Where the number of males exceeds 500 then for every 60	1
(Note: In counting the number of employees, any odd numbers less than 25 or 40 is regarded as 25 or 40)	

must be provided. There are rules concerning the number of conveniences outlined in the Sanitary Accommodation Regulations 1938 as amended in 1974. They are summarised in Table 5.11.

There is an obligation to provide and maintain proper washing facilities including the supply of hot and cold running water. In addition soap and clean towels (or hot air dryers) are also a requirement and must be kept in a clean and accessible condition (see s.58(1) of the Factories Act 1961).

ALCOHOL AND DRUGS

Alcohol has been identified as a contributory factor in a number of accidents in the workplace. It is important that workers are made aware of the dangers of drinking either before going to work or during it. Workers who drink heavily the night before may still be over the legal limit to drive a car the following day. It is unlikely that anyone unfit to drive a vehicle will be fit to operate machinery or equipment at work. An organisation must take steps to include in its policy on health and safety it stance over the use of alcohol, you must consider:

- action to be taken in a one-off situation
- action with a persistent offender.

In the case of the latter it may be possible to offer counselling advice whereas the former might be easily dealt with under a disciplinary investigation depending upon the circumstances. A similar system would be required where workers were found to be dependant on unprescribed drugs.

Workers who are taking prescribed drugs, which medical practitioners have identified as being capable of clouding judgement or affecting manual dexterity, should be afforded some protection and light duties may be a possibility in the short term. Workers should be encouraged to ask their general medical practitioners if any medicines that they have been prescribed will either affect their ability as a road user and as a worker.

STRESS MANAGEMENT

Being able to identify stress in workers must be encouraged. Being able to do something about it is the management function. Stress has long been identified as a contributor to accidents and it is therefore important that it is identified and dealt with. Stress can affect people in different ways. Stress can be eased by relaxing and building confidence, therefore the following are important:

- safety training
- safety education

Table 5.12 Holmes-Rahe scale of life-change units

Event	LCUs
Death of a spouse	100
Marital separation	65
Death of a close family relative	63
Personal injury or illness	53
Marriage	50
Loss of job	47
Marital reconciliation	45
Retirement	45
Change in health of a family member	44
Wife's pregnancy	40
Sex difficulties	39
Gain in family membership	39
Change in financial status	38
Death of a close friend	37
Job change	36
Argument with spouse	35
Taking out mortgage	31
Mortgage foreclosure	30
Change in working responsibilities	30
Sibling leaving home	29
Trouble with in-laws	29
Personal achievement	29
Spouse beginning/stopping work	29
Change in personal habits	24
Trouble with business superior	23
Change in work hours or conditions	20
Change in residence	20
Change in schools	20
Change in recreation	19
Change in social activities	18
Taking out a personal loan	17
Change in sleeping habits	16
Change in family social arrangements	15
Change in eating habits	15
Vacation	13
Minor violation of the law	11

- task analysis and the identification of areas of risk
- bonus schemes and/or time management
- inter-personal relationships
- assertiveness
- career and/or life planning
- counselling
- self-awareness
- negotiation
- team building
- performance management.

However, stress management is best considered within the job profile and recruitment stages and these issues are developed further in other books in this series.

There have been attempts to quantify levels of stress and one such example is given in Table 5.12 referred to as the Holmes-Rahe scale of life change units (LCUs).

CLEANING

There is a requirement under the HSWA for an employer to ensure a safe and healthy working environment. This will require premises to be kept clean and tidy. Most organisations sub-contract this responsibility to office cleaning companies or undertake the task themselves. Office or factory cleaning usually takes place after everyone has gone home, but, as the name implies, cleaning also includes good housekeeping whilst actually at work. Poor housekeeping is responsible for more than half of all work accidents. People should not:

- block or litter walkways/gangways with raw materials or other 'essential' job items
- carelessly discard food or liquids either over floors or equipment
- throw or discard items around the work area which people can trip over
- hang clothing on equipment not designed for the purpose
- cover up important notices or signs
- misuse waste containers
- leave rest rooms and/or kitchen areas littered with motorcycle helmets and other equipment for people to fall over.

HYGIENE

Employers should set out rules for the consumption of food and drink on the premises. Also, those organisations providing works canteens will be subject not only to the HSWA but also the Food Act 1984 and those regulations made under it. It provides for:

- *injurious foods* in that is an offence for any person to add any substance to food, use any substance as an ingredient in the preparation of food or subject food to any other process or treatment so as to render the food injurious to health, with the intent that the food be sold for human consumption in that state
- *protection for purchasers* whereby it is an offence to sell to the prejudice of the purchaser any food which is not of the nature, substance or quality demanded by the purchaser.

It is also an offence to sell food which is unfit for human consumption under section 8(1) of the Food Act 1984.

Where food is kept or prepared, care should be taken to ensure that damaged sink units and refrigerators are well maintained. Personal hygiene should be monitored and provision made for washing and cleaning. Implements used for eating or drinking should not be washed in toilet areas and cracked sinks should be replaced as soon as possible.

ACTION CHECKLIST

- Are your fire procedures adequate?
- When was your last fire drill held?
- Does every employee know what to do in the event of a fire?
- Do your fire marshals know what to do when the fire alarm is sounded?
- Are you satisfied with your building evacuation times?
- Have you sufficient qualified first-aiders?
- Have you sufficient first-aid boxes and are they adequately stocked and checked?
- How often do you check building/room temperatures?
- Is your lighting adequate?
- Is your light bulb replacement programme satisfactory?
- Do you have adequate ventilation in accordance with current regulations and codes of practice?
- Are current radiation requirements relevant to your working situation?

- What procedures do you have in respect of visual display units. For example, do you provide VDU filters and eye-sight checks for certain categories of worker?
- Are your protective clothing policies and procedures adequate?
- For manual handling purposes is your transport policy upheld and maintained?
- Are your storing and stacking guidelines up to date, appropriate and used satisfactorily?
- Do you have appropriate guidelines available for storing and lifting and are these adequate?
- When were your duties and responsibilities under COSHH last examined?
- Are your current procedures, policies and programmes adequate under the COSHH regulations?
- Are your current welfare facilities appropriate and adequate under the requirements of the law?
- Do you have a policy in respect of alcohol and drugs?
- Do your employees know of the dangers of alcohol and drugs and do they know how and where to get help?
- Are your management procedures capable of reducing/keeping stress to a minimum?
- Are your houskeeping and cleaning provisions adequate?

THE OFFICE SAFETY HANDBOOK

'All too often, safety-related decisions are taken which later prove to be too expensive'

'Accidents cost an organisation considerable sums of money'

Finance

INTRODUCTION

Office managers will usually have a basic understanding of management accounting and its importance within the organisational structure. This will allow for decisions to be taken which provide for maximum safety for a known cost. All too often, safety related decisions are taken which later prove to be too expensive or do not provide an appropriate economic rate of return *(ERR)*. Calculating an *ERR* is not be discussed here but those wishing to learn something of this aspect of safety management should read the *Handbook of Safety Management* (Pitman, 1991). Company accountants will need to know the financial implication of all safety management decisions, so it is necessary to spend some time looking at this issue together with some current methods of estimating accident costs and the cost that these incidents can have upon the organisation.

Basically, financial management is a broad term applied to management accounting and funds management. Within the context of management accounting this concerns itself with information that is useful to management. It is defined as the application of accounting knowledge for the purpose of producing and of interpreting accounting and statistical information which is designed to help management in its functions of promoting maximum efficiency and in formulating and co-ordinating future plans and in measuring their execution. This form of accounting practice feeds off financial accounting, cost accounting, budgetary control and capital investment appraisal but the emphasis is on the use of information to assist management to plan and control activities of the organisation rather than upon techniques. Funds management, on the other hand, concerns itself with the acquisition and control of funds; with sources of finance and the control of liquidity. As long-term planning proceeds, financial management must assess the implications of the policies proposed in terms of funds required and make arrangements to raise the long-term funds necessary to secure the financial foundations to provide for organisational growth. Short-term proposals are equally important. Irrespective of how big an organisation is and however profitable it might seem, it is still important to have the money available to pay the bills due and the weekly wages. This emphasises the need to try and forecast future cash requirements both in the short and the long term.

Office managers need to know something about the techniques used to collect and summarise this information but do not need the detailed knowledge that an accountant must have. This chapter discusses basic financial management concepts and does not intend to provide full coverage of a complex subject.

ACCIDENT COSTING

Accidents cost an organisation considerable sums of money and justification for taking positive action within the organisation is based on this fact. As discussed in Chapter 1, your aim will be to reduce accidents or prevent them from happening. Showing remedial strategies to be effective in meeting this aim will in turn save time which can be simply translated into monetary terms. There are many methods for estimating the cost of an accident but there are two classes of costs which result from an accident:

1. from insurance (the insured costs)
2. uninsured costs.

In theory, an organisation can insure itself against any eventuality but in practice the costs of doing this are prohibitive. Most companies, therefore, insure a part of their activities, such as those required by law, and accept liability for the remainder. For example, a worker who drops a casting onto the floor and breaks a toe will have compensation paid from insurance sources. The damage done to equipment or the casting itself may have to be borne by the company if such eventualities are not covered by insurance. Below are 10 elements connected with an incident which may be regarded as uninsured costs. These are not listed in any order of priority:

- cost of wages paid for working time lost by workers who were not injured but whose work output was interrupted by the accident taking place
- the net cost to repair, replace, or straighten up materials or equipment damaged in an accident
- cost of wages paid for working time lost by injured workers, other than workers' compensation payments
- extra costs necessitated by the accident involving overtime work
- cost of supervisors' wages whilst their time is required for activities necessitated by the accident
- wage costs due to decreased output of injured worker after return to work
- cost of learning period of any new worker employed during the injured worker's period of absence

- uninsured medical costs borne by the company
- cost of time spent on accident investigations and processing of HSE requirements and compensation type administration
- additional costs such as equipment replacement, hire or temporary facilities needed until normal state can be resumed. For example, replacement vehicles involved in road traffic accidents.

You must be in a position to calculate your section or department's accident costs and company accountants will usually agree the method to be adopted. A sample form used to extract cost data is given in Figures 6.1 and 6.2.

Total time on light work	☐ ☐ Hours Mins
Employee's average percentage was of normal output during this period	☐ %
Where the injured employee was replaced by additional staff, the wage cost for this period	£ ☐
Time the new employee was below normal for standard wage expected	☐ ☐
New employee's average percentage of expected (normal) output	☐ %
New employee's hourly rate of pay	£ ☐
Time of supervisor in training new employee	☐
Medical cost to the organisation (not covered by employee's compensation insurance)	£ ☐
Cost of time by other clerical workers involved in processing accident paperwork	£ ☐
Other costs not detailed above (e.g. additional costs in renting equipment, public liability etc.)	£ ☐
Total uninsured costs	£ ☐

Figure 6.1 Accident investigator's cost sheet

ABC Manufacturing Ltd

Accident Cost Report

Injury accident number ☐

Non-injury accident number ☐

Dangerous occurence number ☐

Date ☐ Department ☐

Name of injured worker ☐ Works number ☐
(or name of worker reporting a dangerous occurrence)

Please answer the following questions:

1. How many other workers (not injured) lost time because they were talking, watching or helping at the accident? ☐

2. How much time did they lose? ☐☐ Hours ☐☐ Mins

3. How many other workers (not injured) lost time due to damaged equipment in the accident or required the output or assistance of the injured? ☐

4. How much time did most of them lose? ☐☐ Hours ☐☐ Mins

5. Describe the damage to materials and equipment

☐

6. Estimate the cost of repair or replacement of the materials and equipment listed above £ ☐

7. How much time did injured worker lose on the day of injury for which he was paid? £ ☐

8. If operations (or machines) were stopped, will overtime be necessary to make up lost production? YES NO

9. Will it be impossible to make up lost production? YES NO

10. How much supervisor's time was used assisting, investigating, reporting, re-assigning work or instructing a substitute or making adjustments to work loads? ☐☐ Hours ☐☐ Mins

Name of supervisor ☐

Please pass this form as quickly as possible to the personnel manager/manager

Figure 6.2 A supervisor's accident cost report

ABC Manufacturing Ltd

Accident Cost Report

Total lost time (1)	☐ ☐
Treatment time (2)	☐ ☐
First-aid time (3)	☐ ☐
Grand total (1+2+3)	☐ ☐
	Hours Mins

Name of injured worker ☐   Works number ☐

Date of injury ☐ ☐   Nature of injury ☐

Department ☐   Operation ☐   Hourly rate £ ☐

Average hourly rate in the department where the injury occurred £ ☐

Wage cost of lost time by workers not injured (but by employer) £ ☐

Number of workers who lost time because they were talking, watching helping etc. ☐

Average time lost per worker ☐

Number of workers affected by the accident or incident ☐

Average time lost per worker ☐
        Hours Mins

Nature of damage to equipment ☐

Net cost to replace or repair damaged equipment £ ☐

Wage cost of time lost by injured worker while being paid by employer (other than employee's compensation payments) £ ☐

Time lost on the day of injury for which the employee was paid ☐
        Hours Mins

Number of subsequent days absence for which the employee was paid £ ☐

Length of shift or number of hours in standard day ☐
        Hours Mins

Number of additional journeys for medical treatment in employer's time £ ☐

Average time per journey ☐ ☐   Total ☐
    Hours Mins   Hours Mins

Additional lost time by employee for which he/she was paid ☐
        Hours Mins

Lost production costs (include overtime extra supervision etc.) £ ☐

Cost of supervisor's time connected with the accident £ ☐

Supervisor's time shown on report ☐   Additional supervisor's time required ☐
    Hours Mins   Hours Mins

Wage cost due to decrease in output of worker after injury (if rate paid) £ ☐

Figure 6.2 (Continued)

Such data will then allow a company to assess its own average accident costs, which vary from industry to industry, and this will provide for a better basis on which to assess budgetary requirements and improved programme evaluation strategies. It is easier to undertake prospective rather than retrospective studies, although organisations who do not do this may wish to undertake a small retrospective study in order to obtain a sample accident costing. In this case, remember that to help decide how many cases to look at or test the reliability of the averages, compute the standard errors of the averages. It is likely that a two-thirds fiducial probability will give an average for a dozen or more cases of a given type within one standard error of the true average for this type of case in your organisation. The standard error is equal to:

$$\frac{\sigma}{\sqrt{N}}, \text{ and } \sigma = \frac{\sqrt{\Sigma x - \Sigma x X}}{N-1}$$

N = number of cases
x = cost of an individual case
X = average cost

In a larger organisation it might be better to look at 25 to 30 cases.

BUDGETARY CONTROL

Most safety practitioners operate from a budget allocated to them for carrying out company safety policy. Having very briefly discussed how cash is generated for such activities it is now necessary to see how budgets should be controlled. A system of budgetary control establishes various budgets which set out in financial terms the responsibilities of management in relation to the requirements of the overall policy of the organisation. There should be a regular comparison of actual results with budget forecasts both to try and ensure (through action by the safety manager) that the objectives of safety policy are met and also to form a basis for any revision of such policy.

The crux of the budgetary process is that financial limits are *allocated* to component parts of the organisation. Thus, the safety manager plans activities in line with company policy and within the financial limits. It is important to try and obtain departmental management *agreement* to this financial limit as it is the manager who is going to be held responsible for keeping within it. Experience shows that safety managers are more willing to accept responsibility for performance against their budget if their have been allowed to participate in the determination of the size of that budget rather than it being imposed upon them.

Primary responsibility for the administration of the budgetary process is normally delegated by senior management to a budget accountant who has the task of co-ordinating the preparation of both the budgets and their reports. These reports may be presented to a special budget committee (particularly in larger organisations) which is composed of the managers in charge of the major functional areas of the organisation. Membership of this committee would normally be extended to a safety manager.

Preparation of budgets

The following steps are typical of those taken in the preparation of the individual budgets and the *master* budget for a commercial organisation. A safety manager responsible for safety at various locations such as a group safety manager, would be responsible for his/her own master budget. This would then form a part of the overall organisational master budget:

- a statement of overall safety objectives is prepared on which the individual budgets are to be based
- forecasts are made regarding the general economic conditions and the conditions likely to be prevailing in the industry. Here accident data plays a key role
- a safety budget can then be prepared based upon the forecasts and will highlight key task areas for action
- a production budget is prepared in conjunction with the above and will require consideration of all materials and other resources required to carry out those key task areas presented above
- the administrative cost budget is prepared for each area of activity
- a capital expenditure budget covering anticipated changes in legislation or where specialised equipment or modifications is compiled.

The preparation of the budget is shown in Figure 6.3.

Budgetary reports

Once the year has started, the control aspect of the budgetary process consists of comparing the actual results to budgeted figures. The chief value of the budgets as a control mechanism in this connection will be achieved through the effective use of regular reports. These reports, co-ordinated by the budget accountant, will show the variances (normally expressed as percentages) between the actual and budgeted figures. Significant variances should be highlighted. Variance control charts can be employed to present control limits to safety staff and these charts will also show up the significant variances as actual results are recorded.

The reports should also make it clear which budget variances were *controllable* by those responsible and which were not. Detailed explanations of the cause of the variances can be based upon the type of variance analysis between standard unit costs and actual unit costs. It must be remembered that budgetary control needs action.

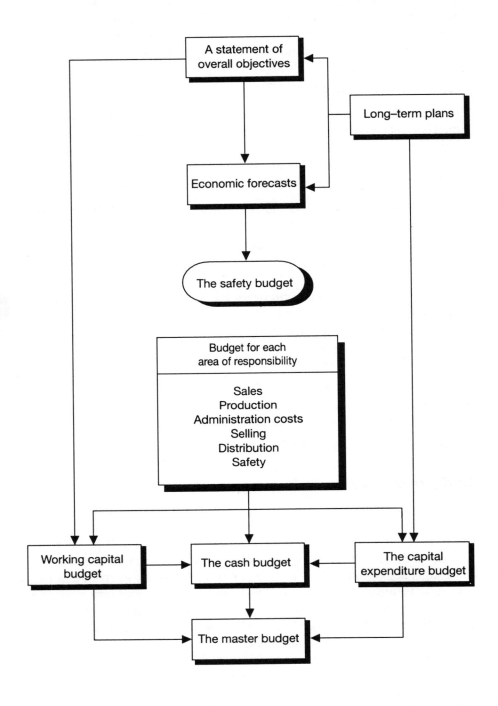

Figure 6.3   The preparation of the individual budgets and the master budget

- reciprocating tools and dies such as power presses and drop stamps
- reciprocating knives, blades and saws
- closing nips between platten motions such as printing machines
- projecting belt fasteners and fast running belts
- nips between connecting rods, rotating wheels, cranks and/or discs
- traps arising from traversing carriages such as metal planing machines.

Whilst there is a vast amount of case law concerning dangerous machinery as well as lists provided by the HSE, it is important for you to:

- identify dangerous equipment
- ensure that dangerous equipment is guarded and/or fenced
- ensure that employees are aware of the dangers
- ensure that appropriate education, training and publicity is given to employees about these dangers
- ensure that all guards/fences are regularly inspected and maintained in good working order.

A checklist for the safeguarding of equipment is given in Table 7.1

Table 7.1 Example list of protection requirements

Guards

- Dangerous parts should be out of reach
- They should be strong enough for the purpose
- Fixed so as not easily removed
- Interlock the guard where necessary so the machine cannot start without the guard being in place
- Have a policy in respect of guards and publicise it to operators
- Seek specialist advice if in doubt
- Machines described as 'dangerous' may only be used by young people when under strict supervision. Some examples are: guillotines, mixers, bacon and vegetable slicers, power wrappers and chain saws

Check that you know how to stop and start the machine

- All fixed guards and safety devices are fixed correctly
- All fixed guards and devices are properly working
- All materials to be used are clear of all mechanical working parts
- The areas around the machine are clear of obstruction and are neat and tidy
- In the event of the machine not working properly there is a recognised and known procedure for reporting and closing machines down
- Appropriate safety clothing and equipment is being worn or used and is serviceable
- Only trained and qualified staff use such machines

CHECKLIST

- Have you listed all the equipment in use in your organisation which is subject to certain legal requirements?
- Have you obtained the various HSE and HSC advice notes concerning the machinery or equipment?
- Are you satisfied with your maintenance programme?
- Are you satisfied that guards and other safety devices are fitted, working and in use as required by company policy?
- Are you satisfied that all your electrical equipment, fittings and appliances conform to laid down safety guidelines?
- When did you last check that your duties and responsibilities described in this chapter are being carried out in a satisfactory manner?

'...it is important to test that administrative procedures ... continue to implement organisational policy efficiently and effectively'

'The aim of the safety audit or review is to improve organisational efficiency'

Auditing safety

WHAT IS SAFETY AUDITING?

There are many interpretations put forward about what exactly safety auditing is. Is it evaluation? Is it monitoring? Is it research? Basically, safety auditing is a means of assessing performance in four areas:

- safety policy
- safety programmes
- safety practice
- safety procedures.

To begin an examination of your performance you should start by answering the following questions which are typical of those asked in management audits. The questions may be listed under the four main heads listed in Figure 8.1.

Consider your activity in each of these areas over, say, the last 12 months and answer these questions.

Inspections

- How many safety inspections have you made?
- How many unsafe conditions were found during these inspections?
- How many of these unsafe conditions were corrected as a result?
- How many unsafe behaviours were observed?
- How many of these behaviours were corrected?
- How many unsafe conditions were reported to you?
- How many of these reports did you respond to?

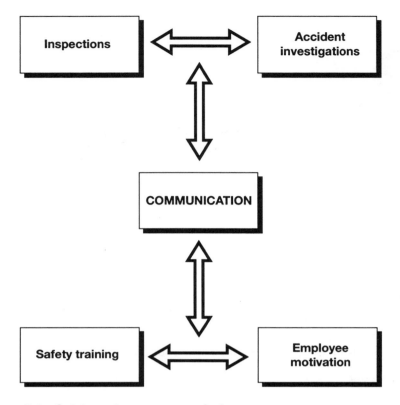

Figure 8.1 Safety performance appraisal

Your answer to these questions will be some indication of your personal involvement in hazard spotting and correcting problems in the area for which you are responsible. You should be taking the lead in the supervision of safety and ensuring that your staff comply with the safety rules.

Inspections should be carried out frequently, how often really depends upon the magnitude of risk involved in the work. Safety representatives are entitled to carry out their own inspections and you should be prepared to respond to their reports appropriately.

Accident investigations

- How many accident investigations did you carry out (in relation to the number of accidents)?
- Were these investigations made promptly or were they made subsequent to an injury benefit claim or safety officer's enquiry?
- How many times did you discover the true cause of the accident? For example, if you ascribed the accident to operator error what factors such as training, supervision, etc. led to this error?

- How many causes could be attributed to failure in the management system?
- How many causes were remedied successfully?

Before choosing to investigate an accident a number of factors are generally taken into account such as the frequency of the occurrence, the severity of the injury, or just the unusual nature of the incident.

In fact there is a duty on all employers under the Social Security (Claims and Payments) Regulations 1979 to take reasonable steps to investigate the circumstances of all accidents notified to them. This investigation should be the starting point in seeking the root causes leading up to the accident itself. If you have relatively few accidents there is no reason why you should not make a serious investigation of them all.

Safety training

- How many new employees were trained in safety matter, for example fire procedures, first-aid facilities, accident or hazard reporting, safety rules, etc?
- How many existing employees have had refresher training in safety matters?
- How many safety briefing sessions have you given or arranged?
- How many of your staff have attended safety training courses?
- What improvements have resulted from this training?

If you found it difficult answer these questions it may be because you have no tangible record of what has been done over the last year. It is important to keep a record of all training done however small, for without this it will be very difficult to demonstrate compliance with the HSWA.

Safety motivation

You have a mission to make effective and efficient use of resources with the maximum degree of health and safety. The concept of efficiency is seen as 'input centred' because it is concerned with ensuring that activities are carried out in a prescribed and appropriate fashion. Effectiveness, on the other hand, is seen as output centred because it is concerned with the extent to which useful achievements are accomplished. An example list is provided in Table 8.1 and you are asked to consider whether the points listed are input or output centred.

The rationale underlying the need for office managers, who have a responsibility for health and safety, to have a clear understanding of their objectives rests on a distinction that can be made between the concepts of efficiency and effectiveness.

Table 8.1 Input or output exercise

Topic	Input centred	Output centred
Internal safety audit (in your experience)		
Ensuring employees arrive/leave work on time		
Expecting tidy work areas		
Appraisal interviews		
Work study exercises		
Objective setting		
Safety budgetary control		
Accident reduction targets		
Expecting staff to be busy and hard working		
Expecting safety equipment/clothing to be worn		
Official personnel procedures		
Safety committee procedures		
The process of identifying safety training needs		
Safety management development courses		
Safety publicity exercises		
Internal safety campaigns (in your experience)		

Effective managers:

- *Do right things*   rather than   *do things right*
- *Produce safe alternatives*   rather than   *accept tidy solutions*
- *Optimise resource use*   rather than   *safeguard some resource*
- *Get results*   rather than   *discharge duties*

Whilst an office manager might be regarded as efficient he/she might not be very effective when output results are measured. It is therefore necessary to set objectives so that areas of efficiency and effectiveness might be appraised within the framework of health and safety requirements both voluntary and statutory. The consequences of not setting realistic objectives would be that an absence of goals will lead to input centred behaviour and role ambiguity and conflict being experienced.

Safety auditing, therefore, must examine objectives and the roles of personnel within its policies, programmes, procedures and practices for efficiency and effectiveness.

SAFETY POLICIES

Health and safety policies have been discussed earlier but policy documents held internally should also include an appendix which clearly outlines the aims and objectives for all staff. An example of the information provided in this document is summarised below:

- an organisational mission statement
- strategic objectives in relation to health and safety
- tactical objectives
- operational objectives.

These *must* be measurable for safety auditing purposes. Resources allocated must also form part of the safety auditing process. This aspect covers:

- legal requirements
- statement of objectives
- budgetary provision
- staff
- organisational structure.

It is also important to question whether the 'safety policy' is adequate. This aspect covers:

- decision making environment
- safety committees
- management involvement
- trades union liaison
- liaison with other interested groups.

Also considered in this section, are attempts to identify areas where difficulties are experienced for remedial action.

AUDITING SAFETY PROCEDURES

In this part of the safety auditing process, it is important to test that the administrative procedures currently in force are able to continue to implement organisational policy efficiently and effectively. Questions relating to this process concern the following issues:

- administrative structure
- communication
- time management
- internal and public relations

- recruitment
- safety training
- supervision
- discipline.

AUDITING SAFETY PRACTICES

This covers those safety practices which may be based upon historical precedence or based upon professional ethics, codes of conduct and practice which directly involve organisational policy and procedure. Questions regarding this concern relate to:

- costing and valuation of accidents
- accident investigation
- data collection
- medical examinations
- welfare
- hazard and risk assessment
- accident analysis
- equipment inspections
- HSE codes of practice
- professional codes of practice.

AUDITING SAFETY PROGRAMMES

Here, the safety audit will examine past remedial strategies for efficiency and effectiveness and involves the four basic elements of the safety mix. These are:

- enforcement
- engineering
- environment
- education.

A sample safety audit form is provided in Figure 8.2. This document is purposely non-specific and is given to illustrate the sort of information which should make up the safety auditing process. There are commercial safety auditing packages available for those who would wish to use them.

Main purpose of job

This contains a brief statement of the reasons for the existence of the job within the organisation. It should not be a description or summary of the activities involved, but should indicate the results or benefits that the job role contributes to the organisation. As a general rule, the main purpose statement should only include one verb, since to have more than one implies that the statement does not describe a single main purpose of the job.

In terms of health and safety a typical safety manager might state that the main purpose of their job is:

> *to prevent accidents from happening*

Key result areas

In every job, there will be a limited number of distinct areas in which effective performance makes a significant impact on the successful achievement of the job purpose. These are referred to as the key result areas.

A key result area is an area of work which is critical to the continued success of the main purpose of the job. Alternatively it is an area where, if the job were not done well, there should be a significant deterioration in the quality or quantity of work and results. One key result area for the safety manager might be:

> *the safety training of all staff*

Key tasks

Within each key result area one or more key tasks can be identified and defined in terms of specific actions and results that those actions are purported to achieve. For example:

- to control..........
- to plan.............
- to determine........
- to develop..........

A key task might be:

> *to ensure that the safety training budget is not overspent*

Standards of performance

Identifying the effect of an activity assists definition of the required level of performance and leads the way to transforming generalised statements of desirable objectives into specific targets against which success can be measured. Standards of performance are statements of the conditions that will exist when the results have been achieved satisfactorily. They should not be set at ideal, but at realistic levels. They must be feasible and be standards against which it is appropriate to judge the employee's performance. One key task may have several performance standards such as:

- *quantity* – how many, how often?
- *quality* – how good, how safe?
- *time* – by when?
- *cost* – at what cost?

Wherever possible, performance standards should be expressed as measurable or objective terms. Some 'good' examples of performance standards are:

- to maintain all safety equipment
- to produce safety literature and advice free from errors in grammar, spelling and punctuation
- to avoid overflowing rubbish bins
- to keep safety equipment service time to not more than 10% of working hours
- to reduce safety maintenance to not more than 15% of working hours from 20% last year.

Some 'bad' examples of performance standards are:

- to maintain safety equipment when requested
- to maintain a high standard of English in safety literature
- to empty waste bins twice a day
- to keep safety equipment serviceable at all times
- to reduce safety maintenance time.

When writing good objectives, an attempt should be made to include as many of the four elements of quality, quantity, cost and time as possible. It is unlikely that a meaningful objective would contain less than two of these elements.

Standards of performance in relation to the safety training (key result area) and financial control of the budget (key task) might be:

- information for the preparation of the annual health and safety budget obtained by (DATE) of each year
- annual estimates prepared by (DATE) each year
- financial control statements examined at the end of each accounting period in order to ensure that:
  - the estimates are revised immediately a new trend becomes apparent
  - the difference between the actual expenditure and estimated expenditure does not exceed x% of the estimate or £y (whichever you have decided).

Control data

The next stage in the process is to identify the information that will be used to assess whether the performance standards are being met and also to identify the documentary sources that will be used to provide this information. This may be obtained from existing control documents, routine statistical information (such as accident data), reports and so on. Control data must be:

- accurate
- available when needed
- available in the right form.

The questions to be asked at this stage are:

- do you have sufficient feedback on progress in the system?
- do you have continuous feedback concerning resource use?
- have you ensured continuous comparison of performance?

In some instances, no adequate control data may be available. In these circumstances one must estimate the time it will take to obtain this information then readjust the criterion accordingly.

Improvement of results

These are the recommendations for changes which are to be made from the safety audit in order that higher standards of performance in key areas is achieved. The types of action that these recommendations are intended to set in motion are:

- improvements in efficiency (resource use)
- changes in procedure or practice
- further safety training, education or publicity
- investigation into problem or high risk areas
- action to improve or change organisational structure.

The questions a safety auditor might ask are:

- what facets of the job cause the most problems, troubles, difficulties and accidents and what needs to be done and by whom, to bring about a worthwhile and lasting change?
- what is currently preventing higher standards being achieved?
- would any of the work be simplified or made more safe if procedures or practice were altered or another section or department did something different?
- what changes, if made, would make tasks easier or less risky to perform?
- where could appreciable savings be made and how?

ACTION CHECKLIST

- *Do you now understand the difference between safety auditing, evaluation, monitoring and reviewing?*
- *Are you satisfied with your current policy in respect of safety auditing?*
- *When was your last safety audit conducted and what happened as a result of it?*
- *Are you satisfied with your current safety policies?*
- *Do your current safety procedures need updating?*
- *Are you satisfied with your current safety programmes?*
- *Are your current safety practices in line with current codes of practice?*
- *Do you ensure that your safety auditor can conduct the exercise with complete independence?*
- *What do you do with the safety audit information which you have received?*

Stevenson College Library
Bankhead Avenue
EDINBURGH EH11 4DE

Useful organisations

Agricultural Training Board
Summit House, Glebe Way, West Wickham, Kent BR4 0RF
Tel: 0181 777 9003

British Medical Association
BMA House, Tavistock Square, London WC1H 9JR
Tel: 0171 387 4499

British Pest Control Association
3 St James' Court, Fire Gate, Derby DE1 1ZU
Tel: 01332 294288

British Safety Council
National Safety Centre, 62–64 Chancellor's Road, London W6 9RS
Tel: 0171 741 1231 (10 lines)

British Standards Institute
2 Park Street, London W1 2BS
Tel: 0171 629 9000

The Chemicals Notification Unit
Central Directorate of Environmental Protection, Department of the Environment, Romney House, 43 Marsham Street, London SW1P 3PY
Tel: 0171 212 3434

Construction Industry Training Board
Bircham Newton, Kings Lynn, Norfolk PE31 6RH
Tel: 01553 776677

The Data Appraisal Unit
Medical Division C5, Health and Safety Executive, Magdalen House, Stanley Precinct, Bootle, Merseyside L20 3QZ
Tel: 0151 951 4000

Faculty of Occupational Medicine
Royal College of Physicians, St Andrew's Place, Regent's Park
London NW1 4LB
Tel: 0171 487 3414

USEFUL ORGANISATIONS

Health and Safety Executive
Daniel House, Trinity Road, Bootle, Merseyside L20 7HE
Tel: 0151 951 4543

(for HSE also see your local area telephone directory)

Health Education Authority
Hamilton House, Mabledon Place, London WC1H 9OX
Tel: 0171 631 0930

Institute of Environmental Health
Chadwick House, Rushworth Street, London SE1 0QT
Tel: 0171 928 6006

Institute of Health Education
14 High Elms Road, Hale Barnes, Cheshire WA15 0HS
Tel: 0161 980 8276/8696

Institute of Home Safety
132 North Road, Dartford DA1 3NB
Tel: 0181 854 8888 ext 8086 (Secretary)

Institute of Occupational Hygiene
132 Oxgangs Road, Edinburgh EH10 7AZ
Tel: 0131 445 1032

Institute of Road Safety Officers
46 Apsley Street, Partick, Glasgow G11 7SW
Tel: 0141 227 2048

Institution of Occupational Safety and Health
222 Uppingham Road, Leicester LE5 0QG
Tel: 0116 2768424

Medical Commission on Accident Prevention
35–43 Lincoln's Inn Fields, London WC2A 3PN
Tel: 0171 242 3176

Road Transport Industry Training Board
Capitol House, Empire Way, Wembley, Middlesex HA9 0NG
Tel: 0181 902 8880

Royal College of Nursing
20 Cavendish Square, London W1M 0AB
Tel: 0171 409 3333

Royal Institute of Public Health and Hygiene
29 Portland Place, London W1N 4 DE
Tel 0171 580 2731

Royal Society for the Prevention of Accidents
Cannon House, The Priory Queensway, Birmingham B4 6BS
Tel: 0121 200 2461

Royal Society of Health
RSH House, 38A St George's Drive, London SW1V 4BH
Tel: 0171 630 0121

Further reading

Other books in this series:

Saunders, *Safety Audit – Designing Effective Strategies* (Pitman), 1992
Saunders, *Handbook of Safety Management* (Pitman), 1991
Stranks, *Safety and Health Handbook* (Pitman), 1992
Croner, *Office Health and Safety*, 1994

RoSPA Health and Safety Series Publications

Index

accident costing	94, 135, 136–40
accident data	92–3, 163
accident details	82, 84, 96–7, 104
accident form	85, 86
accident information	
access to	4, 5
analysis	10, 83, 94
confidential	83, 103
full	104
gathering	1, 3, 4, 9, 81–8, 89–91, 99–105
primary	4, 10, 84, 89–91, 99
published	84, 91
secondary	89, 91
supplementary	4
accident investigation	81–3, 99–105, 152–3
accident locations	87, 94, 97–8
accident numbers	94, 97
accident prevention	83, 105, 158, 161
accident reduction	83, 158
accidents/dangerous occurrences	
investigation and analysis	2, 4, 5, 8, 9, 14, 17, 48, 69, 88
non-injury	87, 94
records	7, 27, 70–1
reduction	2, 8, 10
reporting	47, 50, 70–1, 83, 95
summary sheet	72
time factor	75
see also injury/fatal accidents	
accountability	40–1
accountants	4, 12, 135, 137
accounting, management	135
activities chart	94
additional protection	62
administration	14, 15, 155–6
advisory committees	39, 57
Agricultural Training Board	167
Agricultural (Safety, Health and Welfare Provisions) Act 1956	33
aims and objectives	6, 7, 153–4, 155, 157, 162
air, compressed	50, 51
air quality	116–17, 118
airborne pollution	117
alcohol	98, 129
appeals procedure	45

approved codes 56–7
approved contractors 49
arbitration 77
attendant circumstances
   see contributory factors
Atkin, Lord 26

blame apportionment 103
body corporate 8, 40, 46
breach of duty 27–8, 39, 56
breach of procedure 83
British Medical Association 167
British Pest Control Association 167
British Safety Council (BSC) 5, 7, 166
British Standards Institute 57, 166
budgetary control 6, 135, 140–2, 161, 162
The Building Act 1984 107
Building Regulations 1991 107
building work 51
burden of proof 28, 35

canteens 33, 132
case law 24, 146
casualties 4, 84, 87, 94
casualty information 87, 96–7, 103
casualty, accident-prone 87
centralization 15, 16
certificates
  fire 109–10
  insurance 52, 73
Chemicals Notification Unit 168
children 30, 52
civil service 73
Clapham rail accident 99
Classification, Packaging and Labelling
   of Dangerous Substances
   Regulation 1984 125
cleaning 131
codes of practice 40, 56–7
cognitive interview 103, 104
Commission for the New Towns 73
common law 24, 26, 28, 29–30, 40, 49, 51
company investment 4
company policy (health and safety) 2
comparative studies method 92
compensation 25, 26, 51, 52, 136
complaints 69, 77–8
compliance with rules 48, 78
confidential information 83, 103
Congenital Disabilities (Civil Liability)
   Act 1976 52
Construction Industry Training Board 167

consultants	14, 24
contingency planning	14, 18–19
contract of employment	74
contractors	30–1, 36, 38, 48–51, 66
contributory factors	81, 83, 84, 87–8, 94, 97–8, 103
contributory negligence	29
control data	163
control of premises	40, 41, 51, 52
corporate approach	8
COSHH	125–8
cost-benefit parameters	10, 88
cost-free provision	42, 57, 121
cost restraints	12, 24, 46, 62, 136, 140
costs of accidents	94, 135, 136–40
Criminal Law Act 1977	52
daily management plan	2, 5, 83
damage, accidental	84
damages	33, 52
dangerous occurrences	
see accidents/dangerous occurrences	
see also hazardous …; risk …	
The Dangerous Substances and	
The Control of Substances Hazardous	
to Health (COSHH) Regulations	125–8
data	
analysis	88–9
as management tool	81
base	4, 5, 6, 27, 88, 92
sources	4, 84, 88–9
see also accident data; information	
The Data Appraisal Unit	167–8
death	47, 52, 70, 86–7
decision making	4, 10, 68, 81–2, 84, 92
defective equipment	51
defence pleas	29
delegation of responsibility	5
Department of Social Security	84
directive (EU)	61
directors	
see senior management and directors	
disciplinary action	27, 61, 83–4, 103, 105
discipline, employee	27
discrimination	73–4
diseases	70, 71, 84
dismissal	61, 78
drivers	114, 123
drugs	129
dust and fumes	116–17
duties	
employee	40–1, 46
employer	26, 34, 36–7, 40–1, 74

managerial	1, 3, 9, 23–4, 46–51
manufacturers'/suppliers'	42–3
duty of care	26–30, 121
economic rate of return (ERR)	135
education	1, 5–6, 7, 13, 156
see also training	
effectiveness	154, 156, 158
efficiency	154, 156, 158
electrical equipment	145
electricity	50, 51, 62
Electricity at Work Regulations 1989	145
emissions	40, 116–17, 118
employee awareness	14, 29, 37, 47–8, 51
employee co-operation	99–105, 117–18, 121–2
employee motivation	48, 153–4
employee responsibilities	1, 25, 40–1, 61–2
employer responsibility	26, 34, 40–1, 46, 65–7
Employers' Liability (Compulsory Insurance) Act 1969	52
The Employers Liability (Defective Equipment) Act 1969	51
Employment Acts 1980 and 1982	78
Employment Protection (Consolidation) Act 1978	77
employment relationship	74–5
enforcement	12, 24, 27, 43–6, 81, 105, 146
enforcement notices	44–5
enforcing authority	70, 71
engineering	13, 81, 156
environment	1, 13, 156
see also welfare; workplace	
Equal Pay Act 1970	77
equipment and materials	42–3, 66–7
defective/hazardous	51, 145–8
in use, in accident	84, 98, 99
safety	37, 48, 98, 145–7
unused	50
use of, by others	49
equipment inventory	98, 99
European Union directives	61
evaluation	5, 7–8, 10
excavations	51
experimentation methods	90, 99
expert opinion	61, 90–1
extra-hazardous activities	31, 34
Factories Act 1961	2, 33, 51, 62, 114, 115, 116, 121, 128
Faculty of Occupational Medicine	167
fairness issues	76
Fatal Accidents Act 1976	52
fencing	51

INDEX | **175**

financial management	135–6
fines	39, 43, 45, 52
fire appliances and equipment	108–9, 111
fire authority	107, 109, 110
Fire Precautions Act 1971	2, 107, 109
The Fire Precautions (Factories, Shops and Railway Premises) Order 1989	107
fire procedures/protection	49–50, 62, 66, 107–11
Fire Safety and Safety of Places of Sport Act 1987	107, 109
Fire Service Act 1947	107
fire signing	109
fire warden	109, 111
first aid provisions	62, 66, 112–14
Flixborough Nypro chemical explosion	34
Food Act 1984	132
food and drink	132
firms	70, 81, 84, 85, 86, 94, 95–6, 119, 137–9, 157,
gas	50, 51
general questions	104–5
government departments	5, 84
Government Statistical Office	84
guidance notes	39–40, 125, 127
guilt	46–7
Handbook of Safety Management	135
handling techniques	123–4
hazard sheets	66, 126
hazardous activities	31, 34
hazardous areas	41, 50, 66–7, 69
hazardous equipment	145–8
hazardous substances	40, 66–7, 74, 94, 108, 110, 125–8
hazards assessments	125–8
hazards guidelines	34
health	33, 35, 40, 57, 97
Health and Safety at Work Act 1974 (HSWA)	2, 26, 33–5, 38, 40–3, 44, 45, 56, 65, 74, 77, 107, 108, 114, 115, 116, 117, 121, 128, 131, 132, 153
Section 2:	34, 41, 65–6
Section 37:	46–7
Health and Safety Commission (HSC)	39, 43–4, 56, 57
Health and Safety Executive (HSE)	5, 7, 34, 39, 43–4, 57, 66, 81, 84, 94, 108, 109, 112, 113, 122, 125, 127, 137, 146, 147, 167
Inspectors	40, 44, 45, 67–8, 69
Health and Safety (First-Aid) Regulations 1981	112
Health Education Authority	166
heating	114–15

high risk areas	6
'high status'	83, 102
House of Lords	26
housekeeping in workplace	50–1, 66–7, 110, 130
human error	82–3
hybrid offences	38–9
hygiene	132
hypnosis interview	103
implementation	6, 66–7
improvement notice	44–5, 78
improvements	163
income	74, 75
individual appraisal	158
individual responsibility	46–7
induction course	67
industrial relations	74, 78
industrial tribunals	45, 74–8
information	
collection system	4, 84, 92–9
comparative	4, 91–2
control	163
for staff	23, 37, 38, 41, 47–8, 66–8, 69, 111
sources	2
statistical	4
see also accident information; data	
information technology	5, 8
injunction	26
injury	27–8, 31
injury/fatal accidents	4, 40, 46–7, 50, 52, 70–1, 86–7, 94
injury details	96–7
injury scales	87
injury:time factor	52
input	153–4
inspections	14, 18, 48, 67, 69, 151–2
installers' duties	43
Institute of Environmental Health	166
Institute of Health Education	167
Institute of Home Safety	166
Institute of Occupational Hygiene	166
Institute of Road Safety Officers	166
Institution of Occupational Safety and Health	166
instruction	1, 34, 38, 41
instructors	10
insurance	52, 73, 110, 137
interrogative interview	103
interview techniques	103–4
interviewee	102–3
interviewer 'error'	100, 101
interviews	4, 81, 84, 87, 90, 99, 100–5
job analysis	14, 18

INDEX

job definition	76–7
job evaluation	76–7, 130, 158, 161
key result areas	161, 162, 163
key tasks	161, 162
Law Reform (Contributory Negligence) Act 1945	29
Law Reform (Miscellaneous Provisions) Act 1934	52
leadership	1, 99, 152
legal action	24, 26–7, 39–40
legal expertise	61
legal issues	23
legal precedents	26, 27, 31, 57, 74
legal representation	77
legal right of entry	30
legislation	2, 51–5, 77, 107
see also individual Acts, by name	
legislation awareness	24, 32
'lessons learnt'	61
liability	
civil	30, 51, 146
criminal	31, 33–40, 57
employer	27–31, 51, 69
personal	24, 46–51, 69
vicarious	26, 28
liability insurance	52, 73
life change units (LCUs)	130, 131
lifting equipment	124
lighting	51, 87, 97, 115–16
Limitation Act 1975	52
line manager	99
local authorities	10, 52, 73
London Transport	73
loss distribution principle	51
machinery and safety	145–7
maintenance requirements	23, 34, 145–7
management attitude	12
management failure	46–7
management stategies	8, 26–7
managerial role *see* duties; responsibilities	
manual handling	123–4
manufacturers' duty	42, 66, 146
materials *see* equipment and materials	
Medical Commission on Accident Prevention	167
medical examinations	97
medical suspension	78
medications	98, 129
memory retrieval	103–4
Mines and Quarries Act 1954	33

misuse	41
monitoring	1, 2, 5, 6–8, 12, 49, 62
motivation	48, 153–4
nationalised industries/bodies	52, 73
negligence	27, 28, 29, 31
noise	118–19
noise assessment	118–19, 120
Noise at Work Regulations 1989	117
non employees	23, 30, 38, 73
non injury events	87, 94
notification	70
observation methods	90, 99
occupier	30–1
Occupiers Liability Act 1957	30, 52
offences	38–9, 44, 45–6, 57, 73
Offices, Shops and Railway Premises Act 1963	33, 114, 115, 128
operational reuirements	6
operational systems	7–8
order of events	104
organisational policy and procedure	155, 156
output	153–4
owners (of premises)	40
pencils	98
performance indicators	10
see also safety performance	
performance standards	25, 162, 163
permission requirements	49, 50
physical aspects	37
planning	1, 3, 5, 9, 12, 83
'practicable'	56
police	52
policing	43
precautions	29–30
prejudices	14
premises	30–1, 40
see also workplace	
priority level	2, 6, 10, 24
prison sentence	38–9, 45, 47
prohibition notice	45, 78
prosecution	24, 39–40, 45, 52, 56, 57
Protection of Eyes Regulations	42
protective clothing/equipment	1, 8, 29, 42, 49, 57, 84, 98, 121–2, 146–7
public and work activities	23, 30–1, 40, 41, 50–1
questionnaire	90, 99, 100, 102
questioning methods	90, 99–100, 102
see also interviews	

Race Relations Act 1976	77
radiation-VDUs	119, 121
reasonable care	27–8, 34–8, 46
'reasonably practicable'	56, 57
recall	103, 104
reconstruction of event	104
records	7, 81, 90, 99
records retention	71
recruitment	5, 8, 130
regulations	40, 42, 43, 51, 56–7, 62
see also individually, by name	
relatives/spouse	73
remedial action	4, 5–6, 12, 81, 87, 155
reparation	25
Reporting of Injuries, Diseases and Dangerous Occurrences Regulations 1985 (RIDDOR)	70
research methodology	5, 8, 89
resource allocation	2, 7, 10, 65, 99, 155
resource systems	6
responsibilities	
employers'	2
individual	46–7
levels	41
managerial	23–4, 47–51, 65–7
shared and co-ordinated	4–5, 51
supervisory	99
responsibility accounting	142
review applications	77
review methodology	14–15
RIDDOR	70
risk	
acceptance	29
analysis	27
factors	27, 46
foreseeable	57, 121
inherent	46, 62
management	30, 158
unforeseeable	98
road accidents	94, 102
Road Transport Industry Training Board	167
Royal College of Nursing	168
Royal Institution of Public Health and Hygiene	167
Royal Society for the Prevention of Accidents (RoSPA)	5, 7, 84, 166
Royal Society of Health	166
rule-making processes	75–6
rules	1, 76
safety	33, 34, 40, 57
safety audit	

definition of	12, 151–4
elements of	155–7
framework	158, 159–60
management	158, 63
systems	2
safety committees	10, 43, 67, 68–9
safety education	5–6, 7
safety equipment	37, 48, 98, 145–7
safety inspections	48, 67, 69, 151–2
safety management	8–11, 135, 140–1, 158, 160–1
safety management decisions	81–3
safety management process	11
safety management style	
proactive	5, 6, 27, 63
reactive	26
safety manager's guide	158
safety mix	12, 13, 81, 105, 156
safety motivation	153–4
safety objectives	6, 7, 153–4, 155, 157, 162
safety performance	42, 151, 152, 158, 163
safety policy	
assessment	151, 155
contractors	48–9
document	67, 68, 155
general	65–6
implementation	66–7, 69
written policy	37, 49, 65–7
safety practice	151, 156
safety practitioners	10
safety procedures	83, 84, 151, 155–6
safety programmes evaluation	7–8, 151, 156
safety provision	34, 48
safety publicity	5
safety representatives	6, 37, 43, 67–70, 77, 78, 83, 152
Safety Representatives and Safety Committees Regulations 1977	43, 67
safety review	12, 14–15, 16–17, 19
safety standards	48
safety strategies	62–3, 83
safety systems	23
safety training *see* training	
Sanitary Accommodation Regulations 1938 and 1974	128, 129
Secretary of State	40, 56
security of employment	75
self-certification	87
self-employed	40, 114, 117
self inspection	14, 18
self-recording	99
senior management and directors	2, 3, 10, 15, 46–7, 62, 69
Sex Discrimination Acts 1975 and 1986	73, 77
signs	109

sites, individual	15, 16–17, 41
skills, appropriate	8–9
smoking	50, 109
Social Security (Claims and Payments) Regulations 1979	153
'special projects'	99
stacking/storing	124
staff attitudes	14, 37, 48
staff instruction	23, 34, 37, 48
staff meetings	7
statement interview	102–5
statistics	4, 5, 8
status	83–4, 102
statutory instruments, rules and orders	58–60
statutory law and duty	31, 33, 38, 56
statutory obligations	2, 23–4, 38, 56, 61, 71
statutory protection	33
statutory undertakings	50, 51, 73
stress management	129–30
structural defect	52
structured interviews	101
structured method	6
substantive law	78
supervision	23, 29–30, 34, 37, 41, 48, 49, 51, 99, 103, 152
suppliers' duties	42–3
summary offences	45–6
survey method	88
systems management	6–7
temperature	97, 114–15
testing methods	90
third parties	23, 52, 73
time factor	52, 74, 75, 78
tort	28
Trade Union and Labour Relations Act 1976	77
trade unions	37, 39, 43, 67, 69, 77
training	1, 4, 5, 7–8, 23, 29–30, 34, 37, 38, 41, 66, 69, 77, 78, 110, 113, 123, 127, 129, 153, 161, 162
training courses	39, 97
transport	123
treatment of worker	74, 75
Treaty of Rome	61
trends and objectives	6, 7
Unfair Contract Terms Act 1977	52
unfair dismissal	61
unstructured interviews	101
vandalism	41

ventilation	87, 97, 116–17
vibration	118–19
victim statement	103
visitors	23, 30–1, 41, 66
visual display unit (VDU)	119, 121
warnings	50
water	51, 73
weather conditions	87, 88, 97–8
welfare	33, 34, 40, 128
witness perspectives	104
witness reliability	102, 103
witnesses	81, 84, 102–5
women workers	62, 74
work commencement permission	49
working hours	75
workplace	
housekeeping in	50–1, 66–7, 110
office managers in	15
own or other premises	30
safety	1, 34, 51, 57, 67, 69, 113–14
workstations	94
Zeebrugge ferry disaster	99